Lecture Notes in Computer Science 15235

The series Lecture Notes in Computer Science (LNCS), including its subseries Lecture Notes in Artificial Intelligence (LNAI) and Lecture Notes in Bioinformatics (LNBI), has established itself as a medium for the publication of new developments in computer science and information technology research, teaching, and education.

LNCS enjoys close cooperation with the computer science R & D community, the series counts many renowned academics among its volume editors and paper authors, and collaborates with prestigious societies. Its mission is to serve this international community by providing an invaluable service, mainly focused on the publication of conference and workshop proceedings and postproceedings. LNCS commenced publication in 1973.

Fabio Martinelli · Ruben Rios
Editors

Security and Trust Management

20th International Workshop, STM 2024
Bydgoszcz, Poland, September 19–20, 2024
Proceedings

 Springer

Editors
Fabio Martinelli ⓘ
IIT-CNR
Pisa, Italy

Ruben Rios ⓘ
University of Malaga
Málaga, Spain

ISSN 0302-9743   ISSN 1611-3349 (electronic)
Lecture Notes in Computer Science
ISBN 978-3-031-76370-0   ISBN 978-3-031-76371-7 (eBook)
https://doi.org/10.1007/978-3-031-76371-7

Preface

The International Workshop on Security and Trust Management (STM), from ERCIM's Security and Trust Management Working Group, has established itself as a prominent forum for the discussion of both foundational and applied aspects of security, privacy, and trust management. As we celebrate the 20th anniversary of STM, we are reminded of the significant progress the field has made, and yet, the ongoing challenges that persist.

We are pleased to present the proceedings of STM 2024, held in conjunction with the 29th European Symposium on Research in Computer Security (ESORICS 2024) in Bydgoszcz, Poland. The event continues its longstanding tradition of fostering cutting-edge research and innovative ideas in the realms of security and trust within computing systems.

This edition received 22 submissions, with authors from institutions mainly in Europe, but also from America and Asia. Each contribution underwent a rigorous double-blind peer review process ensuring the high standards of the STM community. After the review phase, a discussion round followed. Finally, only 6 papers were selected as full papers. The quality of submissions allowed us to enrich the event with 4 additional high-quality short papers.

The program was completed with a great keynote by Mauro Conti. Mauro's talk took us through a journey of representative results in the area of side and covert channels. In particular, he discussed attacks such as inferring actions that a user is performing on a smartphone, by eavesdropping on its encrypted network traffic, identifying the presence of a specific user within a network through analysis of energy consumption, inferring information (also key details like passwords and PINs) through timing, acoustic, video, or battery status information, or just the way users play games and listen to the music. An invited talk by Zheng Li entitled "On the Privacy Risks of Machine Learning Models", which was awarded the ERCIM STM Best PhD Award 2024, also complemented the program.

A considerable number of individuals have dedicated a substantial amount of time and effort to guarantee the success of STM 2024. First, we want to thank the Program Committee and the external reviewers for their diligent work in reviewing submissions, which contributed to shape a strong program. Also, we would like to express our gratitude to the people involved in the organization: ESORICS Chairs, Michał Choraś and Sokratis Katsikas; Workshops Chair, Marek Pawlicki; and the constant support of Joaquin Garcia-Alfaro. Special thanks to Pierangela Samarati, chairperson of the ERCIM STM Working Group. Additionally, we want to acknowledge the efforts of Davide Ferraris, who served as Publicity Chair for the event.

And finally, we would like to extend our gratitude to all the authors for sharing their fascinating research findings and to all attendees who joined us and contributed to such engaging discussions. We truly hope that the workshop proceedings will be a valuable source of inspiration and insight for future research in the fascinating field of Security and Trust Management.

September 2024   Fabio Martinelli
                                                                      Ruben Rios

Organization

Program Committee Chairs

Fabio Martinelli   IIT-CNR, Italy
Ruben Rios   University of Málaga, Spain

Publicity Chair

Davide Ferraris   University of Málaga, Spain

Program Committee

Chuadhry Mujeeb Ahmed   Newcastle University, UK
Cristina Alcaraz   University of Málaga, Spain
Mauro Conti   University of Padua, Italy
Sabrina De Capitani di Vimercati   Università degli Studi di Milano, Italy
Carmen Fernandez-Gago   University of Málaga, Spain
Isaac Henderson   Fraunhofer IAO, Germany
Chenglu Jin   Centrum Wiskunde Informatica, The Netherlands
Hiroaki Kikuchi   Meiji University, Japan
Kwok-Yan Lam   NTU University, Singapore
Giovanni Livraga   Università degli Studi di Milano, Italy
Eda Marchetti   ISTI-CNR, Italy
Evengelos Markatos   FORTH, Greece
Weizhi Meng   Technical University of Denmark, Denmark
Marinella Petrocchi   IIT-CNR, Italy
Joachim Posegga   University of Passau, Germany
Kai Rannenberg   Goethe University Frankfurt, Germany
Pierangela Samarati   Università degli Studi di Milano, Italy
Qingni Shen   Peking University, China
Hiroshi Tsunoda   Tohoku Institute of Technology, Japan

Additional Reviewers

Tommaso Bianchi
Giovanni Ciaramella
Davide Ferraris
Alessandro Galeazzi
Manuel Pratelli
Henrich C. Pöhls

Contents

x   Contents

Trust, Anonymity and Identity

DrATC: Dynamic Routing Algorithm Based on Trust Characteristics

Davide Ferraris[1(✉)] and Lorenzo Monti[2]

[1] Network, Information and Computer Security Lab, University of Malaga,
Malaga, Spain
ferraris@uma.es
[2] Cubit Innovation Labs, Via Mario Giuntini, 25, Cascina, Pisa, Italy
lorenzo.monti@cubitlab.com

Abstract. In this paper, we propose a dynamic routing algorithm that leverages various trust characteristics to determine the most trusted path in a network. Trust, a multifaceted concept, encompasses attributes such as direct and indirect experiences, transitivity, directionality, context-dependence, and more. Our approach allows the routing protocol to selectively incorporate these characteristics to enhance the decision-making process. For instance, in scenarios prioritizing direct trust, nodes route packets based solely on direct interactions with their neighbors. In more complex scenarios, both direct and indirect trust are considered, utilizing recommendations from trusted nodes to establish trust with previously un-contacted nodes. We also explore the use of alternative routes based on specific trust values, ensuring sensitive data is routed through the most trustworthy paths. By integrating these trust metrics, the proposed algorithm dynamically adapts to varying network conditions and requirements, improving the overall reliability and security of the data transmission. Our experimental results demonstrate the algorithm's effectiveness in selecting trusted paths and highlight the importance of context and adaptability in trust-based routing. This work contributes to the field by providing a flexible and robust framework for incorporating trust into dynamic routing decisions, paving the way for more secure and reliable network communication.

Keywords: Trust · Network · Routing

1 Introduction

Since the beginning of the human history, it has always been a problem to decide which route to choose either was for hunting or deliver commercial goods. This problem has been transferred in the routing protocols when the first packets were sent to the internet. However, the common denominator in order to choose the path has always been one: trust.

In fact, either we are hunting, deliver a "real" packet or a TCP packet, we need to trust the receiver or the intermediary in order to interact with it. However,

F. Martinelli and R. Rios (Eds.): STM 2024, LNCS 15235, pp. 3–20, 2025.
https://doi.org/10.1007/978-3-031-76371-7_1

if we consider fundamental routing protocols that are still used today in many applications, sometimes trust has been just considered as a default characteristic. For example, for Dijkstra algorithm, it is only important to choose the shortest path from an origin node to a destination node in a network. It is true that the metric chosen can be the distance, time, cost, but we have always to trust the nodes. What happens if one of the nodes in the middle of the transmission wants to behave maliciously? The packet will be lost or worst. Because sometimes, the shortest path is not the best one. This is similar to the children story where the hero has to choose between the right path in a forest where the shortest one was represented by evil trees, fog and bright eyes in the darkness and the longest path was represented by light, peaceful animals and marvelous trees.

Thus, our point is, trust cannot be left out of the equation when choosing a "path". However, trust is difficult to define [7]. There is not a standard definition of it because it is multi disciplinary. Moreover, trust has many characteristics that usually are against each others [1]. Thus, in this paper we will define the characteristics of trust and then we will present a dynamic routing algorithm that is based on the same characteristics.

Internet routing protocols, such as the Border Gateway Protocol (BGP), Open Shortest Path First (OSPF), and Intermediate System to Intermediate System (IS-IS), form the backbone of the internet's infrastructure [14]. They are the algorithms and rules that determine how data packets are directed from one router to another, guiding them along the most efficient and reliable path. These protocols are the unsung heroes of the digital age, silently managing the intricate web of connections that allow us to send emails, stream videos, make online purchases, and access countless online services.

However, our task is design each routing algorithm as they will only be based on the trust levels among the nodes. Such nodes can be an Internet of Things (IoT) device belonging to a cluster, sensors in a common network, basically everything that can be connected. Thus, we will only focus on trust and general nodes. We will represent a decentralized network where each node can have enough computational power to compute a trust value according to the different characteristics [26].

The paper is composed as follows. In Sect. 2, we discuss the related work. In Sect. 3, we present the trust characteristics. In Sect. 4, we present the dynamic routing algorithm based on trust characteristics. In Sect. 5, we will propose three examples showing how the algorithm work and then, in Sect. 6, a use case presenting on how the trusted routing algorithm works. Finally, in Sect. 7, we conclude and present the future work.

2   Related Works

In this section, we will firstly present definitions of trust, then we will discuss about which algorithm exists in routing and finally we discuss about existing works about trusted routing.

2.1   Trust

Trust is a multifaceted concept with varied definitions across disciplines [18]. For this reason, trust is a foundational concept in many fields, including sociology, psychology, and computer science [3]. In the context of computer science, trust refers to the confidence in the reliability, integrity, and security of entities within a system, such as nodes in a network, software components, or entire systems [8]. The concept of trust is particularly critical in decentralized and distributed systems, where direct oversight and control are limited.

Several definitions of trust have been proposed in the literature, each emphasizing different aspects of the concept. We describe three of the many of them:

- Marsh [18]: Marsh introduced the idea of formalizing trust as a computational concept, defining it as a value that can be used to predict the future behavior of an entity based on past interactions. This approach laid the groundwork for incorporating trust into automated decision-making systems.
- Gambetta [11]: Trust is a means to reduce the complexity of interactions in an uncertain environment, serving as a mechanism to manage the uncertainty associated with the actions of others. Gambetta's definition underscores the role of trust in simplifying complex, uncertain interactions.
- Jøsang [15]: Trust is the subjective probability by which an entity believes that another entity will perform a particular action on which its welfare depends. This definition highlights the probabilistic and subjective nature of trust.

In computer science, trust is often modeled and quantified to enhance the security, reliability, and performance of systems [13]. Trust models are used in various domains, including network security, distributed systems, and e-commerce [9].

Trust models in computer science typically involve the following components [10]:

- **Direct Trust**: Derived from direct interactions and experiences with an entity. For example, if a node in a network consistently forwards data packets correctly, it gains a higher direct trust score.
- **Indirect Trust (Reputation)**: Based on recommendations or observations from other entities. If multiple nodes report positive interactions with a particular node, that node's reputation (indirect trust) increases.
- **Hybrid Trust**: Combines both direct and indirect trust to form a more comprehensive trust evaluation. This approach helps mitigate the limitations of relying solely on one type of trust metric. Trust metrics can be computed using various mathematical and probabilistic methods, including Bayesian networks, fuzzy logic, and weighted averages.

However, several applications of trust have been developed in computer science.

One of them has been performed for Ad-Hoc and Sensor Networks. In these decentralized networks, trust-based routing algorithms are used to ensure secure and reliable communication. For example, Ilyas et al. [14] proposed a trust-based routing framework for ad-hoc networks that isolates malicious nodes by evaluating trust metrics.

Trust has also been implemented in Peer-to-Peer (P2P) Networks [4]. Here, trust models help identify reliable peers and mitigate the risk of malicious behavior. Systems like BitTorrent use reputation systems to encourage cooperative behavior among peers.

Trust is even more important considering also the final users in E-Commerce systems where trust plays a crucial role in online transactions [15]. Systems like eBay and Amazon use trust and reputation systems to build consumer confidence and reduce the likelihood of fraud.

Moreover, we can state that in Cloud Computing, trust management in cloud environments ensures that users can rely on cloud service providers to handle their data securely and reliably. Trust models assess the trustworthiness of different cloud services based on factors like service history and security practices [22].

However, we believe that trust can play a crucial role if it is mainly considered in routing algorithms. For this reason, we will describe now what are routing algorithms and then the existing routing algorithms that consider trust partially.

2.2   Routing Algorithms

Routing algorithms are essential components in network systems, responsible for determining the optimal paths for data transmission from source to destination [14]. Traditional routing algorithms can be broadly classified into two categories: static and dynamic.

On one hand, Static Routing Algorithms involve predefined routes that do not change unless manually reconfigured. Examples include algorithms used in simple, small-scale networks where changes are infrequent. Dijkstra's algorithm is one of them [6]. This is a graph search algorithm used to find the shortest path from a starting node to all other nodes in a weighted graph. It uses a priority queue to repeatedly select the node with the smallest known distance, updates the shortest paths to its neighboring nodes, and marks it as visited. The process continues until the shortest paths to all nodes are determined. This algorithm guarantees the shortest path in graphs with non-negative edge weights [2].

On the other hand, Dynamic Routing Algorithms adapt to network conditions in real-time, responding to changes in network topology, traffic load, and link failures [14]. Common examples include:

- Distance Vector Routing (DVR): Each router maintains a table (vector) of the minimum distance to every other router. The Bellman-Ford algorithm is a foundational approach in DVR.
- Link State Routing (LSR): Routers have complete network topology information and independently compute the shortest path to every other router using algorithms like Dijkstra's, but in a dynamic way.

- Path Vector Protocols: Used in inter-domain routing, such as the Border Gateway Protocol (BGP), which maintains path information that gets updated as routes change.

2.3   Trusted Routing Algorithms

Trusted routing algorithms incorporate trust metrics into the route selection process to enhance security, reliability, and performance. These algorithms extend traditional routing methods by integrating trust evaluations of nodes and links, considering factors such as past behavior, recommendations, and security credentials.

Trust-Based Routing in Ad-Hoc Networks is useful. In fact, Ad-hoc networks, due to their decentralized nature and lack of fixed infrastructure, particularly benefit from trust-based routing. Pirzada et al. [25] proposed a trust-based routing framework for ad-hoc networks, where trust metrics are used to identify and isolate malicious nodes, thereby improving network security and reliability.

In Wireless Sensor Networks (WSNs), trust-based routing algorithms aim to ensure data integrity and network longevity. Khan et al. [17] presented a trust-aware routing protocol that evaluates trustworthiness based on direct and indirect observations, thus enhancing the resilience of the network against attacks and failures.

Some approaches integrate trust mechanisms into existing routing protocols. For instance, Perkins [24] enhanced the Ad hoc On-Demand Distance Vector (AODV) protocol by incorporating a trust model that assesses node reliability based on historical interactions, thereby improving route selection and network performance.

However, we found lack of trust considerations in all of these methods. In fact, we believe that in order to fully consider trust in routing algorithms, we should consider that trust metrics can be computed using various methods, including:

- Direct Trust: Based on direct interactions and experiences.
- Indirect Trust: Derived from recommendations or observations from other nodes.
- Hybrid Approaches: Combine both direct and indirect trust metrics to form a comprehensive trust evaluation.
- Other Trust Characteristics: see the next section.

For this reason, we will now present the characteristics of trust and around them we will build the trusted algorithm. In order, to dynamic choose them when computing a routing path.

3   Trust Characteristics

In this paper, we propose a routing algorithm enhancing by trust considering its characteristics.

We have identified several characteristics proposed by different authors during the years. We summarize them in Table 1. In the first column there is the trust characteristic and in the second one the paper mentioning such characteristic.

Table 1. Characteristics of trust

Direct	[3]
Indirect	[1]
Transitive	[5, 27]
Directed	[27]
Dynamic	[4, 12, 23]
Context-dependent	[1, 20]
Local	[1, 5]
Global	[1]
Specific	[16, 19]
General	[16, 19]
Asymmetric	[21]
Subjective	[12, 27]
Objective	[1]
Composite-property	[12, 27]
Measurable	[27]

Now, we specify trust characteristics describing the meaning of each of them:

1. **Direct.** Trust is based on the direct experience. We can also say that trust depends on past history [3].
2. **Indirect.** Usually, if direct experience is absent, we can start computing a trust value considering the recommendation of other entities [1]. This is the basis of systems based on reputation.
3. **Transitive.** Trust can be also considered as transitive [5]. In fact, trust can be conditionally transferable, as there is the possibility to transmit/receive trust information through a path of recommendations [27].
4. **Directed.** Trust is directed. It means that we have an oriented relationship between different entities [27]. Thus, it is possible that if an entity A trusts an entity B, the opposite can be not the same (i.e., B distrusts A).
5. **Dynamic.** Trust change over time, it can increase or decrease due a several actions. Chang [4] states that "trust builds with time". In fact, an entity could trust another entity for a determined context in a specific moment, but this can change positively or negatively in a future moment [23]. Moreover, as

Grandison stated [12] "trust must be able to adapt to the context in which a trust decision has been made and can change according to different contexts".

6. **Context-dependent.** As we mentioned before, trust is strictly connected to the context. "In general, trust is a subjective belief about an entity in a particular context [27]." and more specifically "where the trust of a node i in a node j varies from one context to another [1]".

7. **Local.** Trust can be **local** [1] because it depends on a couple of entities (i.e., Alice and Bob) and if we consider other two couples (i.e., Alice and Charlie, and Bob and Charlie), it is possible that Alice distrust Charlie, even if Bob trusts Charlie [5].

8. **Global.** As Abdelghani stated "trust also called reputation means that every node has a unique trust value in the network which can be known by all other nodes [1]".

9. **Specific.** On the one hand, we can state that trust can be specific [16,19]. This happens because an entity can trust another entity only for a specific purpose or service.

10. **General.** On the other hand, trust can be considered as general [16,19]. In this case, the an entity A trusts an entity B independently from the purpose or a specific context.

11. **Asymmetric.** This means that two entities tied by a relationship may differently trust each other. It means that even if A trusts B, this does not imply that B trusts A [21].

12. **Subjective.** Trust is subjective because it is related to a personal opinion based on different factors (i.e., past experience) and these factors can be differently important for different entities [12]. In fact, trust is perceived in a dissimilar way for each individual in a particular context [27].

13. **Objective.** Trust can be also considered **objective** "such as when trust is computed based on Quality of Service (QoS) properties of a device [1]". Furthermore, an objective parameter to compute trust is also known as **reputation**. Connected to *indirect*.

14. **Composite-property.** Trust can be composed of different attributes. For example as Grandison [12] stated it can be composed of "reliability, dependability, honesty, truthfulness, security, competence, and timeliness". Thus, compositionality is an important aspect for trust computations [27] and every attribute could have different weight.

15. **Measurable.** Finally, trust is measurable. In fact, "trust values can be used to represent the different degrees of trust an entity may have in another. [27]." This characteristic is the basis for the computation of a final trust value during trust management.

The aforementioned characteristics and their relationships are explained in Fig. 1.

The outer circle indicates that the characteristics listed there are always present. The traits within the inner circle remain significant in all contexts (i.e., directed and asymmetric). *Transitive* is italicized because it is not always applicable and is placed in a separate rectangle. Additionally, three pairs of characteristics are connected by dotted lines, signifying that they are mutually exclusive.

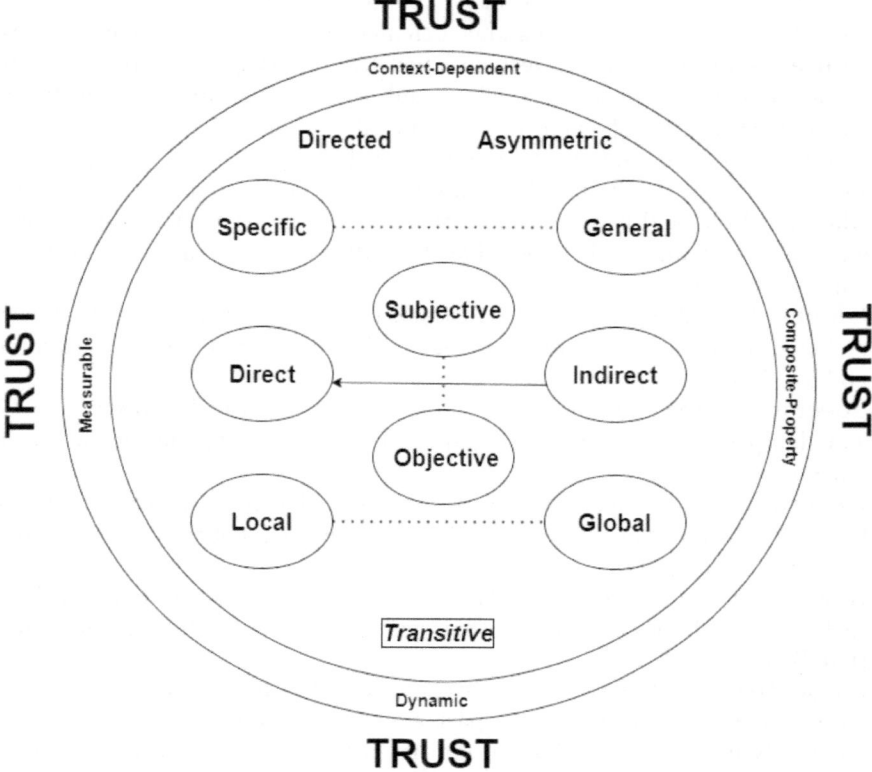

Fig. 1. Trust Characteristics and Their Relationships [9]

Specifically, trust can be either specific or general, subjective or objective, and local or global. However, trust can also simultaneously be specific, objective, and global.

At the center of the diagram is the pair: direct and indirect. An arrow extends from indirect to direct, indicating that indirect trust can sometimes lead to the formation of direct trust. This occurs when there is no prior direct knowledge (i.e., no past interactions), and an indirect parameter is needed to start building a trust value. This process is illustrated by the arrow.

4   DrATC: Dynamic Routing Algorithm Based on Trust Characteristics

As we mentioned before, dynamic routing algorithms are essential for maintaining efficient and reliable communication in networks, particularly in environments where network conditions and node trustworthiness can vary over time. By incorporating trust characteristics into dynamic routing algorithms, we can enhance the security and reliability of the network. This section outlines a general model for a dynamic routing protocol that leverages trust metrics to determine the most trusted paths for data transmission.

The trust-based dynamic routing algorithm integrates traditional routing metrics (i.e., hop count, latency) with trust characteristics to make more informed routing decisions. The algorithm continuously evaluates and updates trust levels of nodes based on their behavior and interactions, ensuring that the most trusted paths are chosen dynamically as network conditions evolve.

In Sect. 3, we have discussed about trust characteristics. For our algorithm, we want to take mainly into considerations two aspects. Direct trust and indirect trust. For the latter, we can also consider transitive trust.

More specifically:

- Direct Trust: Nodes evaluate direct trust based on past interactions. This involves recording and analyzing the outcomes of previous data exchanges, such as successful transmissions and detected security breaches.
- Indirect Trust: Nodes consider recommendations from other trusted nodes. This mechanism allows nodes to gather trust information about nodes they have not interacted with directly. If we consider transitive trust, nodes can infer trust relationships through chains of trusted nodes, allowing them to build trust with nodes beyond their immediate neighbors.

However, we will consider other important trust characteristics into the routing metrics, such as reliability, security, competence, and context-specific trust values. These metrics are weighted and combined to form a composite trust score for each node.

As trust is dynamics, we want to transfer this capability to the algorithm too. For this reason, trust levels are dynamically updated based on ongoing interactions and feedback from other nodes. This ensures that trust evaluations reflect the most current network conditions and node behaviors. Thus, the protocol adjusts trust scores in real-time, considering factors such as recent successful transmissions, detected anomalies, and recommendations from other nodes.

As also context is fundamental for trust considerations, we perform trust evaluations depending on the context, meaning trust scores can vary based on the type of data being transmitted and the specific requirements of the communication (i.e., higher security needed for sensitive data). Nodes dynamically adjust trust evaluations based on the context, ensuring that routing decisions align with the specific needs of the network at any given time.

In the algorithm, we implement a Routing Decision Process (RDP) considering the following steps: Path Discovery, Path Evaluation and Path Selection. More specifically:

1. Path Discovery: Nodes initiate path discovery processes to identify potential routes to the destination. During this process, nodes exchange trust information and routing metrics.

2. Path Evaluation: Each potential path is evaluated based on its overall trust score, which is a composite of the trust scores of the intermediate nodes along the path.
3. Path Selection: The path with the highest trust score is selected for data transmission. This ensures that the chosen path not only meets traditional routing criteria but also maximizes trustworthiness.

However, as the algorithm is dynamic, feedback and learning are necessary. Thus, nodes continuously monitor the performance of the selected paths and provide feedback to update trust evaluations. This feedback loop allows the protocol to learn from network conditions and improve routing decisions over time. Nodes share their trust evaluations and experiences with other nodes, enhancing the overall trust awareness and cooperation within the network.

The flow related to the paths and related feedbacks is showed in Fig. 2.

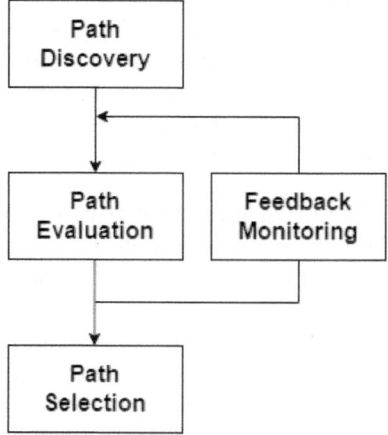

Fig. 2. Algorithm Flow and Feedback

4.1 General Operation of the Model

In this part, we describe the general behaviour of the algorithm. Firstly there is the initialization phase.

So, each node initializes its trust evaluations based on direct experiences and recommendations from other nodes. Nodes also set up mechanisms to dynamically update trust scores as interactions occur. Path Discovery and Trust Evaluation:

When a node needs to send data, it initiates a path discovery process, during which it gathers trust information about potential intermediate nodes. Nodes exchange routing and trust information, allowing the initiating node to evaluate the trustworthiness of each potential path.

The initiating node evaluates the trust scores of all potential paths and selects the one with the highest composite trust score. The selected path is used for data transmission, ensuring that the most trusted nodes are involved in the communication.

During and after data transmission, nodes monitor the performance of the selected path and provide feedback to update trust evaluations. Trust scores are dynamically adjusted based on the success or failure of the transmission, recommendations from other nodes, and any detected security issues.

The algorithm continuously adapts to changing network conditions and node behaviors. Trust evaluations are updated in real-time, ensuring that routing decisions remain optimal and secure.

5 Algorithm Utilization Scenarios

As we presented in the previous sections, trust has several characteristics. We want to enable our routing algorithm in order to choose among them in order to compute trust according to the selected one. For example, in this section we show a routing algorithm considering only direct trust, without consider transitive or indirect trust. In this case, each node will have to choose among different nodes only considering the direct experience.

In the second example, we will provide the possibility to enable both direct and indirect, in order to have the possibility to compute a trusted value also for nodes with no direct experience.

The third one, will consider the possibility that a trust value changes after receiving a feedback and dynamically change the more trusted route.

In all of these examples we will consider trust as directed as described in Sect. 3.

5.1 Scenario 1: Direct Trust-Based Routing

In this scenario, the routing protocol is designed to consider only direct trust. Each node in the network maintains a record of its direct interactions with neighboring nodes. The trust value is calculated based solely on these direct interactions. When a node needs to forward a packet, it selects the next hop based on the highest direct trust value. This ensures that the path chosen consists of nodes that have had positive direct interactions with their immediate neighbors.

We consider four different nodes and the need to be satisfied is that Node A needs to send packets to Node D.

In this scenario, we have the following situation also represented in Fig. 3:

- Node A has direct trust values for Node B (0.9) and Node C (0.7).
- Node B has a direct trust value for Node D (0.8).
- Node C has a direct trust value for Node D (0.6).

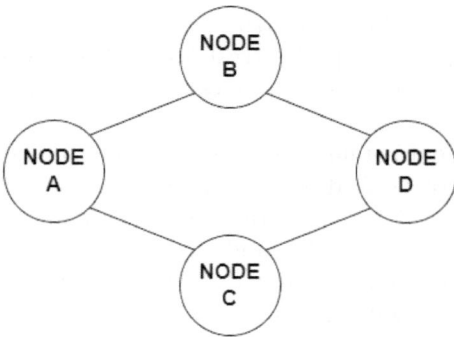

Fig. 3. Nodes Distribution in Scenario 1

According to these values, the routing decision will be the following:

Node A selects Node B as the next hop because the direct trust value (0.9) is higher than for Node C (0.7). Then, Node B forwards the packet to Node D based on its direct trust value (0.8).

In this case, there are no feedback changing the trust values.

5.2   Scenario 2: Direct and Indirect Trust-Based Routing

In this scenario, the routing protocol is enhanced to consider both direct and indirect trust. Nodes can rely on recommendations from trusted neighbors to establish trust values for nodes with which they have no direct interactions. This approach allows the network to be more flexible and robust, especially in dynamic environments where direct interactions may be limited.

In this case we have six nodes and the need is for Node A to send packets to Node F.

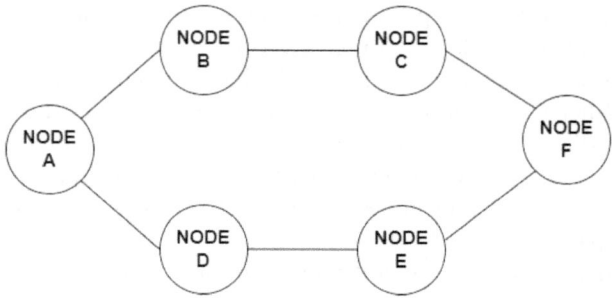

Fig. 4. Nodes Distribution in Scenario 2

In this scenario, we have the following situation also represented in Fig. 4:

- Node A has direct trust values for Node B (0.8) and Node D (0.7).
- Node B has a direct trust value for Node C (0.3).
- Node D has a direct trust value for Node E (0.9).
- Node C and Node E have a direct trust value for Node F and they are (0.85) for C and (0.85) for E.

According to these values, the routing decision will be the following:

Node A selects Node B as the next hop because the direct trust value (0.8) is higher than for Node D (0.7). However, the indirect trust value communicate by B to A according to the following step C is dramatically lower than the indirect trust value communicated by D to A according to Node E: (0.3) vs (0.9). So, the algorithm prefers to proceed from A through D instead of B. Then, the final values from C and E to F are the same (0.85). Thus, the final path will be A -> D -> E -> F.

5.3   Scenario 3: Adapting Routes Based on Feedback

In this scenario, we consider also the dynamic feedback mechanism to adjust trust levels in real-time. In this scenario, a node initially trusted for routing behaves maliciously, compromising the trust relationship. Our protocol must then adapt by recalculating and finding a new trusted path based on updated trust values.

Let's consider a network with nodes A, B, C, D, and E. The initial trust values between these nodes are as follows:

- Node A to Node B: 0.9
- Node A to Node C: 0.7
- Node B to Node D: 0.8
- Node C to Node D: 0.85
- Node D to Node E: 0.9
- Node B to Node E: 0.75
- Node C to Node E: 0.8

The goal is to find the most trusted path from Node A to Node E. Initially, the protocol selects the path based on the highest trust values. The most trusted path is A -> B (0.9), B -> D (0.8) and D -> E (0.9).

The total trust value for this path is $0.9 * 0.8 * 0.9 = 0.648$.

However, suppose Node A receives feedback indicating that Node B has behaved maliciously, and its trust value drops to 0.4. The updated trust values are:

- Node A to Node B: 0.4
- Node A to Node C: 0.7
- Node B to Node D: 0.8
- Node C to Node D: 0.85
- Node D to Node E: 0.9
- Node B to Node E: 0.75

– Node C to Node E: 0.8

With the updated values, the protocol must recalculate the path from Node A to Node E, prioritizing nodes with higher trust values. The recalculated path is: A -> C (0.7), C -> D (0.85) and D -> E (0.9)

The total trust value for this new path is 0.7 * 0.85 * 0.9 = 0.5355.

Although the new path's total trust value is lower than the initial path, it avoids the malicious behavior of Node B, ensuring a more secure route. This dynamic adjustment mechanism helps maintain the integrity and security of the data transmission, even in the face of changing network behaviors.

In all of these examples, trust is considered directed, meaning that the trust relationship from one node to another is not necessarily reciprocal. This ensures that the routing decisions are based on the specific trust values in the given direction, optimizing the trustworthiness of the chosen path.

6 Use Case: Trust-Based Routing in a Network

In the previous section, we have presented three scenarios in which we have showed how the algorithm works for direct and indirect trust and what happens when a trust value change over time. In this section, we present a more complex use case considering different trust characteristics to find the most trusted path from point A to point B incorporating a variety of trust metrics into the routing algorithm.

In this use case, node A wants to send data to node B. The network comprises several intermediate nodes (e.g., C, D, E, F, etc.), each with varying levels of trustworthiness based on different trust characteristics. Node A must select the most trusted path to ensure data integrity and security.

The different options are the following:

1. Direct Trust: Node A has previous interactions with node C and has a history-based trust value for node C.
2. Indirect Trust: Node A has no direct interaction with node D but receives recommendations from node C about D.
3. Transitive Trust: Trust can be passed along a chain, e.g., if A trusts C and C trusts D, then A can consider trusting D through C.
4. Directed Trust: Trust is not reciprocal, meaning A might trust C, but C might not necessarily trust A.
5. Dynamic Trust: Trust levels change over time; A may update its trust value for C based on recent interactions.
6. Context-Dependent Trust: Trust values vary depending on the context (e.g., trust for data transmission might differ from trust for control messages).
7. Local Trust: Trust relationships are specific to pairs of nodes; A might trust C differently than it trusts D.
8. Global Trust: Each node has a reputation score known to all nodes in the network.

9. Specific Trust: A may trust C for routing control messages but not for data transmission.
10. General Trust: A may generally trust D without specific context.
11. Asymmetric Trust: Trust between A and C may not be mutual.
12. Subjective Trust: Trust evaluations are subjective and based on A's perspective.
13. Objective Trust: Trust is computed using objective metrics like QoS or uptime.
14. Composite-Property Trust: Trust evaluations consider multiple attributes like reliability, competence, and security.
15. Measurable Trust: Trust values are quantifiable, allowing for comparison between paths.

The algorithm, will proceed with the steps presented in Sect. 4.

In the Path Discovery step, Node A gathers trust information about all intermediate nodes (C, D, E, F, etc.) from its direct interactions, recommendations from other nodes, and global reputation scores.

Then, during the Path Evaluation, we can have four different approaches. The first one is related to the Direct Trust computation, in which the algorithm calculate direct trust scores based on A's history with intermediate nodes. Another option is to consider Indirect Trust, thus there is incorporate recommendations from trusted nodes. Another possibility is to compute Transitive Trust scores using known trust relationships. Another option, is to consider only Context-Dependent Trust. In this case, there will be an adjustment of trust scores based on the specific context of the data transmission.

After choosing the approach, Node A evaluates possible paths to B considering the aggregated trust scores of intermediate nodes. Each path is assigned a trust score based on the composite trust values of the nodes in the path.

Now, there is the Path Selection part where the path with the highest trust score is selected as the most trusted path from A to B. This path should ideally maximize direct and indirect trust, consider transitive trust relationships, and account for the context and dynamics of trust values.

After the path is chosen, it is possible to proceed with Data Transmission. Thus, Node A sends the data to node B via the selected trusted path. During transmission, trust levels are monitored and adjusted dynamically based on the ongoing interactions and feedback.

An example is the following. Assume the following trust values:

- Direct Trust (A -> C): 0.8
- Indirect Trust (A -> D via C): 0.7 (C recommends D)
- Transitive Trust (A -> E via C and D): 0.6 (A trusts C, C trusts D, D trusts E)
- Context-Dependent Trust: Adjusted based on the type of data (i.e., more stringent for sensitive data)
- Composite Trust: Consider factors like reliability (0.8), security (0.9), and competence (0.7)

The trust calculation must be performed in order to calculate the trust score for each path:

– Path 1 (A -> C -> B): Trust Score = 0.8 (Direct Trust with C)
– Path 2 (A -> D -> B): Trust Score = 0.7 (Indirect Trust with D)
– Path 3 (A -> C -> D -> B): Trust Score = 0.7 * 0.8 = 0.56 (Transitive Trust via C and D)
– Path 4 (A -> E -> B): Trust Score = 0.6 (Transitive Trust via E)

Based on these calculations, Path 1 (A -> C -> B) would be chosen as the most trusted path.

In this example, Node A utilizes various trust characteristics to evaluate and select the most trusted path to Node B. By integrating direct, indirect, transitive, and other trust properties, the routing decision is optimized to ensure the highest level of trustworthiness, adapting dynamically to changes in the network. This approach highlights the complexity and importance of trust in secure and reliable network routing.

7 Conclusion and Future Work

In this paper, we proposed a dynamic routing algorithm that leverages trust characteristics to determine the most reliable paths in a network. By incorporating various trust metrics such as direct, indirect, and transitive trust, our approach allows for a nuanced evaluation of node relationships, ensuring that routing decisions are made based on comprehensive trust assessments. The flexibility to enable or disable specific trust characteristics demonstrates the adaptability of our protocol to different network scenarios and security requirements.

As a future work, we will improve the algorithm considering all the trust characteristics, defining it in a more rigorous way and applying it to complex scenarios. Moreover, we will propose an adaptation of BGP including trust characteristics.

Acknowledgments. This work has been supported by the EU project HORIZON-MSCA-2021-SE-01 under grant agreement No 101086308 (DUCA) and SecTwin 5.0 (TED2021-129830BI00) funded by the Ministry of Science and Innovation (Agencia Estatal de Investigación (AEI)/10.13039/501100011033) and by the European Union "NextGenerationEU"/PRTR.

This work reflects only the authors view and the Research Executive Agency is not responsible for any use that may be made of the information it contains.

References

1. Abdelghani, W., Zayani, C.A., Amous, I., Sèdes, F.: Trust management in social internet of things: a survey. In: Dwivedi, Y.K., et al. (eds.) I3E 2016. LNCS, vol. 9844, pp. 430–441. Springer, Cham (2016). https://doi.org/10.1007/978-3-319-45234-0_39
2. Agudo, I., Fernandez-Gago, C., Lopez, J.: A model for trust metrics analysis. In: Furnell, S., Katsikas, S.K., Lioy, A. (eds.) TrustBus 2008. LNCS, vol. 5185, pp. 28–37. Springer, Heidelberg (2008). https://doi.org/10.1007/978-3-540-85735-8_4
3. Beth, T., Borcherding, M., Klein, B.: Valuation of trust in open networks. In: Gollmann, D. (ed.) ESORICS 1994. LNCS, vol. 875, pp. 1–18. Springer, Heidelberg (1994). https://doi.org/10.1007/3-540-58618-0_53
4. Chang, J., Wang, H., Gang, Y.: A dynamic trust metric for p2p systems. In: 2006 Fifth International Conference on Grid and Cooperative Computing Workshops, pp. 117–120. IEEE (2006)
5. Christianson, B., Harbison, W.S.: Why isn't trust transitive? In: Lomas, M. (ed.) Security Protocols 1996. LNCS, vol. 1189, pp. 171–176. Springer, Heidelberg (1997). https://doi.org/10.1007/3-540-62494-5_16
6. Dijkstra, E.: Dijkstra's algorithm. http://en.wikipedia.org/wiki/Dijkstra_algorithm. Accessed 12 Oct 2007
7. Fernandez-Gago, C., Moyano, F., Lopez, J.: Modelling trust dynamics in the internet of things. Inf. Sci. **396**, 72–82 (2017)
8. Ferraris, D., Fernandez-Gago, C., Lopez, J.: A trust by design framework for the internet of things. In: NTMS'2018 - Security Track (NTMS 2018 Security Track), Paris, France (2018)
9. Ferraris, D., Fernandez-Gago, C., Roman, R., Lopez, J.: A survey on IoT trust model frameworks. J. Supercomput. **80**(6), 8259–8296 (2024)
10. Fortino, G., Fotia, L., Messina, F., Rosaci, D., Sarné, G.M.L.: Trust and reputation in the internet of things: state-of-the-art and research challenges. IEEE Access **8**, 60117–60125 (2020)
11. Gambetta, D., et al.: Can we trust trust. In: Trust: Making and Breaking Cooperative Relations, vol. 13, pp. 213–237 (2000)
12. Grandison, T., Sloman, M.: A survey of trust in internet applications. IEEE Commun. Surv. Tutor. **3**(4), 2–16 (2000)
13. Hoffman, L.J., Lawson-Jenkins, K., Blum, J.: Trust beyond security: an expanded trust model. Commun. ACM **49**(7), 94–101 (2006)
14. Ilyas, M., et al.: Trust-based energy-efficient routing protocol for internet of things–based sensor networks. Int. J. Distrib. Sens. Netw. **16**(10), 1550147720964358 (2020)
15. Jøsang, A., Ismail, R., Boyd, C.: A survey of trust and reputation systems for online service provision. Decis. Support Syst. **43**(2), 618–644 (2007)
16. Kenning, P.: The influence of general trust and specific trust on buying behaviour. Int. J. Retail Distrib. Manag. **36**(6), 461–476 (2008)
17. Khan, T., Singh, K.: TASRP: a trust aware secure routing protocol for wireless sensor networks. Int. J. Innov. Comput. Appl. **12**(2–3), 108–122 (2021)
18. Marsh, S.P.: Formalising trust as a computational concept. Ph.D. thesis, Department of Computing Science and Mathematics, University of Stirling (1994)
19. Morrow Jr., J.L., Hansen, M.H., Pearson, A.W.: The cognitive and affective antecedents of general trust within cooperative organizations. J. Manag. Issues, 48–64 (2004)

20. Moyano, F., Fernandez-Gago, C., Lopez, J.: A conceptual framework for trust models. In: Fischer-Hübner, S., Katsikas, S., Quirchmayr, G. (eds.) TrustBus 2012. LNCS, vol. 7449, pp. 93–104. Springer, Heidelberg (2012). https://doi.org/10.1007/978-3-642-32287-7_8
21. Nitti, M., Girau, R., Atzori, L.: Trustworthiness management in the social internet of things. IEEE Trans. Knowl. Data Eng. 26(5), 1253–1266 (2014)
22. Nuñez, D., Fernández-Gago, C., Luna, J.: Eliciting metrics for accountability of cloud systems. Comput. Secur. 62, 149–164 (2016)
23. Pavlidis, M.: Designing for trust. In: CAiSE (Doctoral Consortium), pp. 3–14 (2011)
24. Perkins, C.E.: Ad hoc on-demand distance vector (AODV) routing, internet-draft. draft-ietf-manet-aodv08.tx (2001)
25. Pirzada, A.A., McDonald, C., et al.: Establishing trust in pure ad-hoc networks. In: ACSC, vol. 4, pp. 1. Citeseer (2004)
26. Roman, R., Najera, P., Lopez, J.: Securing the internet of things. Computer 44(9), 51–58 (2011)
27. Yan, Z., Holtmanns, S.: Trust Modeling and Management: From Social Trust to Digital Trust, pp. 290–323. IGI Global, Hershey (2008)

Distributed Dynamic Self-control Anonymity Management Model

Alperen Aksoy[1]($\boxtimes$)(iD) and Dogan Kesdogan[2](iD)

[1] Friedrich-Alexander University of Erlangen-Nuremberg, Erlangen, Germany
alperen.aksoy@fau.de
[2] University of Regensburg, Regensburg, Germany
dogan.kesdogan@ur.de

Abstract. In this study, we present a dynamic anonymity management model for communication networks that allows users to self-control
their desired level of anonymity protection. Inspired by well-established
closed-loop control systems, which continuously adjust parameters based
on feedback, we incorporate similar feedback mechanisms into anonymous communication systems. With this continuous feedback mechanism, users can measure their current anonymity protection and take
action (e.g., sending fake messages) to increase if their protection level is
below than desired. Thereby, we distribute the task of anonymity protection between the user and the system. The experimental results demonstrate that the proposed model effectively achieves and maintains the
desired level of anonymity protection.

Keywords: anonymity · management · traffic analysis · statistical
disclosure attacks · mix networks

1 Introduction

In today's digital landscape, the widespread adoption of the Internet brings
about security and privacy challenges for users regarding their data. In communication services such as e-mail and instant messaging, while encryption techniques
protect the content of messages, it is important to note that important traffic
data, such as the sender and recipients of messages, remain unprotected from
network observers (e.g. Internet service providers). This traffic data is valuable
and reveals significant details about communication patterns and social connections. For example, a person with a life-threatening disease may repeatedly
contact a specific hospital department, risking rejection of health insurance application if the company obtains this traffic information. Because cryptographic
techniques cannot solve the problem of protecting the confidentiality of these
data, anonymity protection methods address this problem. Information-theoretic
anonymity protection (i.e., zero information flow) requires perfect coordination
among users to use network services (e.g., email, web browsing) simultaneously.

F. Martinelli and R. Rios (Eds.): STM 2024, LNCS 15235, pp. 21–35, 2025.
https://doi.org/10.1007/978-3-031-76371-7_2

Since this perfect coordination is not feasible in practice (i.e., the Internet), solutions are focused on achieving strong practical anonymity protection like Loopix Protocol [22].

Consequently, previous models and systems have proposed adjustable system parameters to achieve the desired protection level (e.g., average message delay, batch size, message-sending schedules) [2, 7, 10–12, 17, 22, 23, 25]. However, their settings are generic, i.e., they may not be sufficient to achieve the personally desired level of protection. In addition, these techniques are static, lacking dynamic methods for users to take action as required based on the current threat. Therefore, there is a need for a flexible approach on the user side to dynamically take action (e.g., sending additional dummy messages) to maintain the desired level of protection.

In this paper, we propose a dynamic anonymity protection management model that allows users to self-control and maintain their desired level of protection in open network environments (e.g. the Internet). We distribute the anonymity protection task between the user and anonymous communication system. To the best of our knowledge, we are the first to propose such a dynamic anonymity management model in communication networks.

The remainder of the paper is structured as follows: Sect. 2 covers related works on anonymity protection in communication networks, while the proposed model is presented in Sect. 3. The details and results of the experimental studies are presented in Sect. 4. The results are discussed in Sect. 5 and the conclusions and future works are given in Sect. 6.

2   Related Works

The protecting confidentiality of network traffic data first addressed by Chaum with the introduction of Mix networks[1]. To achieve perfect protection of these data via Mix technique, Chaum stated that all users must contribute to all rounds with an equal number of messages, and these messages should be broadcasted to all users in encrypted format [4]. Such coordination was possible at the time (early 1980s) because the communication networks only had a few thousand users. However, in today's open environments such as the Internet with billions of users, this approach is no longer practicable.

Therefore, first-generation anonymous communication networks such as Mixmaster [21] and Mixminion [5] were released without user coordination, i.e., users could continue their free behavior to join communication rounds. As a result, it has been shown that these techniques are vulnerable to long-term observation attacks [1, 6, 16]. Subsequently, several approaches were proposed that follow Chaum's idea for a fixed small group of participants and oblige every user to take part in every communication round. In these approaches, if users do not have a real message to send for a specific round, they must send a dummy message

---

[1] Readers who are not familiar with the MIX technique are provided with a simple model in Sect. 3.1 under **Mix networks**, which contains the information required to understand our work.

to maintain perfect coordination [12,17,24]. However, since these systems are deployed on the Internet, where people are free to shut down the client and go offline, they are still vulnerable to long-term observation attacks due to lack of coordination caused by offline users.

Since the anonymity protection relies on the persistent anonymity sets (see Fig. 2 for user coordination), there are approaches presented to burden this task to the anonymous communication system. In [25], when Alice sends her first message, the system records other participants of her first round as buddies of Alice. In future rounds, Alice's messages will be delayed until at least k buddies send a message, before forwarded to the recipient. This approach prevents users from utilizing the system if their buddies offline or leave the system. In addition, this approach has the fundamental weakness that a delay in the messages by an attacker is not noticed by Alice and all users, as the delay can also come from the approach (i.e., a statistical analysis can benefit from the delay of individual messages).

Addressing this issue, Hayes et al. [11] groups users based on their message-sending behavior. They suggest delaying Alice's messages until other group members send their messages to the anonymous communication network. If all users in the group do not send messages within a pre-determined waiting time, the present messages are then forwarded to the recipients. Since the message-sending behavior of users may change over time, this approach is also vulnerable to long-term observation attacks. These methods take a step forward from previous approaches, providing stronger anonymity protection in open environments. However, they bring significant overhead to the anonymity system, as it must perform calculations for each user in every communication round, and they can also cause significant latency for message delivery.

There are also approaches proposed to enhance anonymity protection by increasing user coordination to use network services simultaneously. These methods suggest dividing users into groups based on their message-sending behavior. In [10], a message-sending schedule is assigned to groups, requiring users to follow their group's schedule to send messages at the same time. However, this static approach has a major drawback: if users miss their scheduled rounds due to issues (e.g. connection interrupts, electricity cuts), they can be de-anonymized by adversaries in a few steps. Lastly, [2] suggests assigning a message-sending rate for each group. Users send messages following to a Poisson process, according to their group's message-sending rate to coordinate participation in rounds. While this approach helps to synchronize users, it does not establish perfect coordination. Although this method can extend the time before Alice is de-anonymized, it cannot prevent the identification of all her communication partners.

Previous approaches to enhance anonymity protection have been designed statically and generic for all users. That is, they are not under the control of Alice (i.e., the users) as they are centralized and therefore more or less non-transparent for Alice. Thus, do not comply with the highest principle of privacy-the principle of self-determination. They furthermore either impose significant overhead on the anonymous communication system by requiring it to calculate the current protection status for each user or assign strict schedules to maintain user coordination. In addition, they lack a dynamic mechanism that allows users to assess their current protection status during communication.

3   User Centered Dynamic Anonymity Management Model

In this paper, we propose a dynamic anonymity management model that enables users to self-control their level of anonymity protection, inspired by the well-known *closed-loop control* system model in control engineering. A closed-loop control system is a mechanism that continuously adjusts its output based on feedback to achieve a desired performance or setpoint. We adapt this model for anonymity protection in communication networks. An overview of the model is shown in Fig. 1.

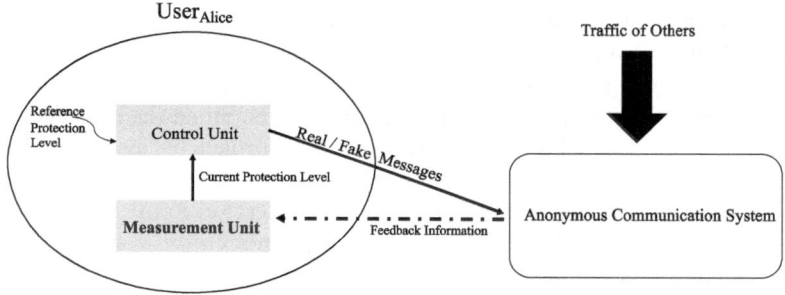

Fig. 1. Overview of the proposed self-control anonymity management model

In the figure, the user model contains two main components: measurement and control units. The measurement unit periodically collects feedback information (e.g., previous recipients, number of active users) from the anonymous communication system to measure the current protection level for the user. It uses sensitive measurement methods to evaluate the current protection level and then forwards the measurement results to the control unit. The control unit compares the measurement results with the desired level of protection (reference protection level) and takes the necessary actions to maintain this.

With this model, users can be aware of their current anonymity protection level and take actions to increase it if necessary. This model distributes the task of protecting anonymity between the user and the anonymous communication system. The system provides feedback information to users, allowing them to calculate their current level of protection and take self-actions as needed. Since the system does not take additional actions to increase the privacy protection for each user, this method also ensures that the user's privacy considerations are not known by the system (e.g., which communication contacts Alice specifically tries to hide).

In this model, when Alice wants to send a message to one of her communication partners, she prepares the message and adds it to her message buffer (located on her computer). The measurement and control units then work collaboratively to emit this message from the buffer and send it to the anonymous communication network to be forwarded to the respective recipients. In this section, we will demonstrate in detail the functions of each component in the model over our reference implementation.

3.1 Measurement Unit

The measurement unit performs a critical task, because the responses of the control unit to adjust the security level rely on accurate vulnerability measurements. To measure the vulnerability (i.e., protection level) of anonymous communication systems, researchers have revealed several de-anonymization attacks [1,3,6,16]. Since the proposed model utilizes dummy messages (fake messages sent to randomly selected users to confuse the adversary) to enhance anonymity protection, this technique effectively thwarts exact type of traffic analysis attacks, which rely on combinatorial analyses of collected anonymity sets. As a result, the adversary is limited to obtaining information through the statistical analysis of these sets. Therefore, attacks that utilize statistical analyses become a useful option for measuring anonymity protection, i.e., assessing the adversary's current information gain.

A well-known example is the Statistical Disclosure Attack (SDA), introduced by Danezis to determine the contact list of targeted users based on statistical communication properties [6]. In the SDA, the adversary observes the network by recording a list of users who receive messages when the targeted user (Alice) sends a message. Since other users who send a message at the same time as Alice change over time due to their different system usage behaviors, the attacker expects Alice's messaging partners to appear more frequently than others in the observed recipient lists. To demonstrate the effectiveness of the measurement unit, we utilize the SDA to measure the current protection level in our reference implementation because of the following advantages: (i) The attack always provides results even after the first observation i.e., users can be aware of how far the adversary to correctly detect their contacts at each step. This gives them the opportunity to take action before being de-anonymized. (ii) The calculation of the attack consists of only a few vector operations and does not bring significant overhead to users. (iii) The application area of the attack is broad, meaning it can be applied to any type of anonymous communication system with complex embedding methods (e.g., Pool-Mix [8]) and the presence of additional traffic from dummy messages.

To demonstrate the SDA, Mix networks [4] are used that operate in discrete rounds, during which incoming messages are shuffled before being forwarded at the end of each round. We also consider that the anonymous communication system works as a mix network to simply demonstrate our ideas in reference implementation. As a user action to increase the current protection level, we utilize end-to-end dummy messages, the effectiveness of which was demonstrated in [18].

Mix Networks presented by Chaum to prevent any arriving message from being linked to a departing one through special routers called mixes [4]. The formal model of the mix technique is illustrated in Fig. 2. We consider a system in which N users communicate via the mix network. In each communication round, a mix collects a batch of b (batch size) messages from users in set A, cryptographically transforms them, and then forwards to the corresponding recipients in set B in a random order. Thus, the relationship between incoming and outgoing messages is concealed. Subset $A' \subset A$ denotes the senders who have sent messages in the current mixing round (i.e. the sender anonymity set), subset $B' \subset B$ the recipients who receive any of these messages (i.e. the recipient anonymity set). Mix networks can be structured as cascades of mixes in a layered network topology. This configuration ensures that no single mix node has access to the complete sender-recipient information, and strengthening the anonymity sets. However, (statistical) disclosure attacks treat anonymous communication systems as black boxes and only observe incoming and outgoing traffic on the network. Since this architectural approach alone cannot mitigate this threat, we assume that Mix networks operate as a single mix in our reference implementation.

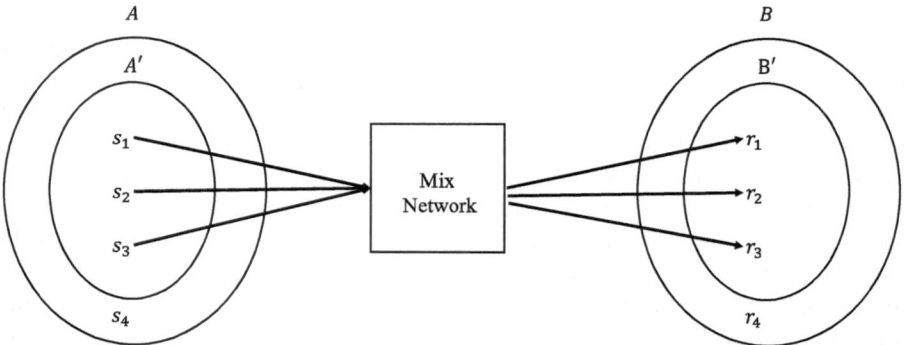

Fig. 2. Overview of communication via mix network

The Statistical Disclosure Attack (SDA): As noted above for the sake of simplicity, the SDA follows the model shown in Fig. 2. The adversary observes the mix network and records an observation (i.e. the according recipient set $|B'| = b$) from each communication round in which Alice sends a message. Each observation consists of a batch of b messages, where 1 message is from Alice, considered as the *signal*, and the remaining $b - 1$ messages are from other users, considered as the *background noise*. The goal of the SDA is to statistically filter out the background noise in order to detect Alice's overall signal (peer set of

Alice) from the noisy channel, utilizing a sufficient number of observations. It is further assumed and parameterized with m that Alice communicates via the mix network with one of her m communication partners in each round. Following the example in above Fig. 2 and assuming that s_1 is Alice then one of the recipients r_1 to r_3 is peer of Alice (i.e., the recipient) and the others are noise.

Let $\vec{v}$ be a vector with N elements corresponding to each user in the system, representing Alice's communication behavior. Assuming Alice has m communication partners and sends messages uniformly to her partners; the values of m elements are expected to be $\frac{1}{m}$, while the remaining elements are 0. Also, $\vec{u}$ is a vector with N elements, denotes the message-sending behavior of users other than Alice, and represents noise estimation. Noise is assumed to be uniform across all users. Therefore, all elements of $\vec{u}$ are set to $\frac{1}{N}$.

Let $i = 1$ to t, where t denotes the number of communication rounds observed by the adversary. Each element of $\vec{o_i}$ corresponds to a user and denotes the probability that the user is the recipient of a message sent by Alice in round i.

Based on the law of large numbers, within a large set of observations, the arithmetic mean of the observations gives [6]:

$$\overline{O} = \frac{1}{t} \sum_{i=1}^{t} \vec{o_i} \approx \frac{\vec{v} + (b-1)\vec{u}}{b} \tag{1}$$

This information can be used to determine Alice's communication behavior $(\vec{v})$. From Eq. 1, $\vec{v}$ can be derived as:

$$\vec{v} \approx b \frac{\sum_{i=1}^{t} \vec{o_i}}{t} - (b-1)\vec{u} \tag{2}$$

Evaluation of the Attack Results: In de-anonymization attacks, the well-known metric to evaluate vulnerability is the number of required obervations to complete the attack, i.e., correctly identifying partners of Alice. However, there is no clear endpoint for the SDA. Therefore, to measure the vulnerability of systems by the SDA, this metric is commonly used to calculate the number of observations required to identify with high probability the known partners of the targeted user applying the attack. The result vector $\vec{v}$ reflects the probabilities of the users that can be identified as Alice's peer communication partners from the observation and calculation. Thus, the highest m most probable users are considered as identified communication partners of Alice by the attack.

Feedback Information from the System: As noted above, to simply demonstrate our ideas, we consider the anonymous communication system working as a Mix Network. We only add one property to the original model: at the end of each round, the round's observation vector $(\vec{o_i})$ will be returned to the participating senders as feedback information. Thereby, all users use this vector to measure their current protection level (e.g., the communication partners those at risk of identifying by the SDA).

In Fig. 3, we demonstrate the functionality of the measurement unit with a small example. In this example, there are three senders (s_1, s_2, s_3) each communicating with their respective communication partner. In each round, the mix forwards incoming messages to respective recipients after collecting two messages, i.e., the batch size (b) is set to 2. The relationships between senders and recipients are $(s_1 \rightarrow r_1, s_2 \rightarrow r_2, s_3 \rightarrow r_3)$. We designate Alice as sender s_1. In the figure, two communication rounds that Alice participates in are shown.

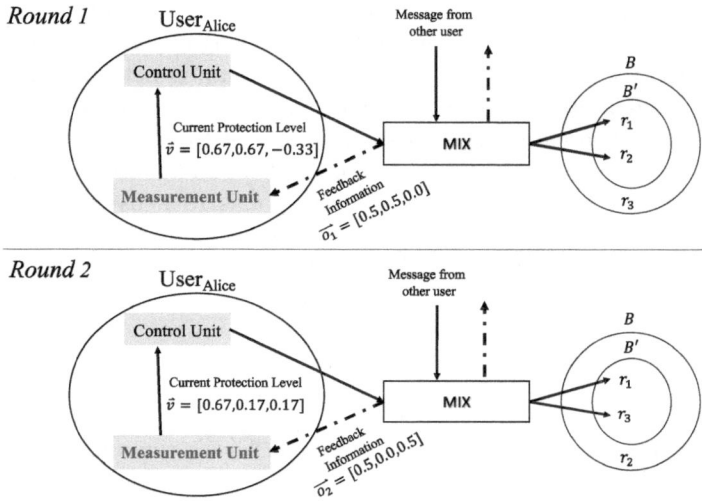

Fig. 3. A small example communication scenario following the proposed model

In the first round, recipients of forwarded messages are $\langle r_1, r_2 \rangle$. The observation vector of the round is calculated as $\vec{o_1} = [0.5, 0.5, 0.0]$ by the mix, and this information is forwarded to the participating senders (s_1 and s_2). Alice's measurement unit calculates the attack results following Eq. 2, $\vec{v} \approx b.\vec{o_1} - (b-1).\vec{u} = [0.67, 0.67, -0.33]$. The vector $\vec{v}$ denotes Alice's statistical communication behavior as calculated by the adversary targeting to de-anonymize her. Considering the current $\vec{v}$, the adversary assumes that r_1 or r_2 is equally likely to be Alice's communication partner. Evaluating this result is a task for the control unit, details of which will be given in the next section.

In the second round, incoming messages are forwarded to the recipients $\langle r_1, r_3 \rangle$. After forwarding messages to the respective recipients, the mix sends feedback information to the participating senders (s_1 and s_3) as $\vec{o_2} = [0.5, 0.0, 0.5]$. Alice's measurement unit then calculates the current $\vec{v} \approx b \cdot \frac{\vec{o_1} + \vec{o_2}}{2} - (b-1) \cdot \vec{u} = [0.67, 0.17, 0.17]$. As seen in the current $\vec{v}$, the adversary assumes s_1 is communicating with r_1 with significantly higher probability than other recipients. This result provides information about Alice's current anonymity protection. Deciding whether the current anonymity protection is sufficient for Alice's privacy considerations is the task of the control unit.

Since the senders of the round are not included in the feedback information, Alice can only calculate the attack results targeting herself. Therefore, our model does not give users the opportunity to switch to the attacker role.

3.2   Control Unit

The control unit is responsible for embedding Alice's messages into the anonymous communication system. These messages can be real or dummy based on the current protection level. If the current measurement results indicate that the actual MIX technique provides higher protection than the reference level, a real message emitted from the buffer is sent. On the contrary, if the provided protection level is lower than desired, a dummy message is sent to a randomly selected user[2] to increase the current protection level. This approach corrupts the above described statistics of the SDA and confuses the adversary. The working principle of the control unit is shown in Algorithm 1.

Algorithm 1 Pseudo-code of the control unit

//This function will be called after each message added to the user buffer.
function PROCESS_MESSAGE
    //The control unit requests the current protection level from the measurement unit.
    $current\_protection\_level \leftarrow request\_current\_measurement()$
    if $current\_protection\_level > reference\_protection\_level$ **then**
        $send\_next\_message\_from\_buffer()$ //send a real message
    else
        $send\_dummy\_message()$ // sending blank message to random user
        $PROCESS\_MESSAGE()$ // recall the function because there is a message in the buffer
    end if

end function

---

4   Experimental Study

We conducted a series of experiments via simulations to evaluate the effectiveness of the proposed approach. We developed a simulation program in Python, utilizing the Simpy package [20] to simulate message-sending events of users and mix nodes.

Firstly, we applied the SDA over a constant number of rounds targeting to Alice, a user in an anonymous communication system follows the proposed

---

[2] We assume that implicit information is added to dummy messages, enabling only the recipient to efficiently recognize and delete them with minimal effort. A corresponding technique is detailed in [9].

approach. We tracked Alice's protection (i.e., detection rate of communication partners) varying the referenced anonymity protection level. Then, we investigated the effect of system parameters (batch size, number of users, number of communication partners) on the efficient utilization (ratio of sent real messages compared to dummy messages) of users to the proposed system. During the simulations, we followed common assumptions as used in previous studies [13, 19]; each user has m communication partners randomly selected from N users. They send messages according to a Poisson process with the same message-sending rate and uniformly select the recipient of each message from their own recipient set. The provided simulation results are averaged over 50 repetitions and confidence interval was set to 0.95.

4.1 Anonymity Protection as Desired During Communication

In this section, we evaluate the proposed model applying the SDA to observe Alice's anonymity protection during communication rounds. We set the system parameters with the batch size (b) as 50, the number of users (N) as 1000, and the number of communication partners (m) for each user as 10. To evaluate the measurement results, we assume that the control unit compares the highest m probability users (set of detected contacts by the attack) in calculated vector $\vec{v}$ (see Sect. 3.1) with the set of Alice's actual communication partners. The ratio of the intersection of these two sets is considered as the correct detection rate. Alice's policy considers her privacy protection sufficient until the ratio of correctly detected partners by the adversary reaches the referenced detection rate (RDR). We varied the RDR, from 20% to 80%, and also included a scenario without additional protection (RDR: -). We observed 1000 rounds from the mix network for each simulation and recorded the rate of detected partners of Alice after each observation. The results of simulations are shown in Fig. 4.

According to the results, if users do not adopt an additional approach to protect their anonymity, their communication partners are quickly identified by the adversary using the SDA. When users adopt the proposed anonymity protection approach, they can maintain their anonymity close to their referenced anonymity protection degree (RDR). To maintain this desired protection level during communication rounds, users must send dummy messages instead of real messages in some rounds. If achieving the desired protection level is inefficient under the current system settings, this may cause utilization issues, limiting users' ability to send their real messages to recipients. In the next section, we will consider the utilization of users under different system settings.

4.2 Utilization

In this section, we investigate the cost of maintaining the desired anonymity protection by the proposed approach in terms of utilization. The excessive use of dummy messages may prevent users from efficiently utilizing the network to send real messages. We consider the ratio of real messages compared to dummy

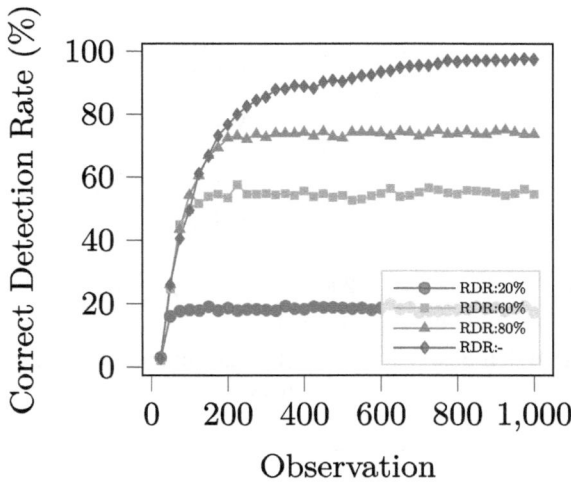

Fig. 4. The SDA results varying the reference protection level (RDR)

messages as a utilization metric (see also [14]). The utilization for Alice (ρ_{Alice}) can be calculated as:

$$\rho_{Alice} = \frac{\text{number of real messages sent}}{\text{total number of messages sent}} \quad (3)$$

We evaluate the utilization of Alice under the different system settings via simulations. We set the default system parameters with the batch size (b) as 50, the number of users (N) as 1000, each user has 10 communication partners (m), and the targeted reference detection rate of partners (RDR) as 20%. We varied one parameter for each experiment while keeping the others fixed according to the default settings. We observed 1000 rounds from the mix networks for each simulation and calculated the utilization value for Alice (ρ_{Alice}) in each scenario. The results are shown in Fig. 5.

The utilization of the anonymous communication system by Alice increases with the batch size (see Fig. 5-a). A smaller batch size (b) scenarios requires to sending more dummy messages to maintain the reference protection level. Alice's utilization decreases with an increasing number of users (N) in the system. Therefore, we suggest scaling up the batch size proportionally with the number of users in the system for efficient utilization. Additionally, utilization increases with the number of communication partners (m) because detecting more users requires additional observations, thereby reducing the need for dummy messages.

The utilization of Alice decreases with the increasing reference protection level (i.e., decreasing RDR), meaning that maintaining higher protection requires the use of more dummy messages. Previous studies show that increasing the batch size and the number of communication partners increases the number of observations required to correctly identify Alice's communication partners. Conversely, increasing the number of users decreases the required observations

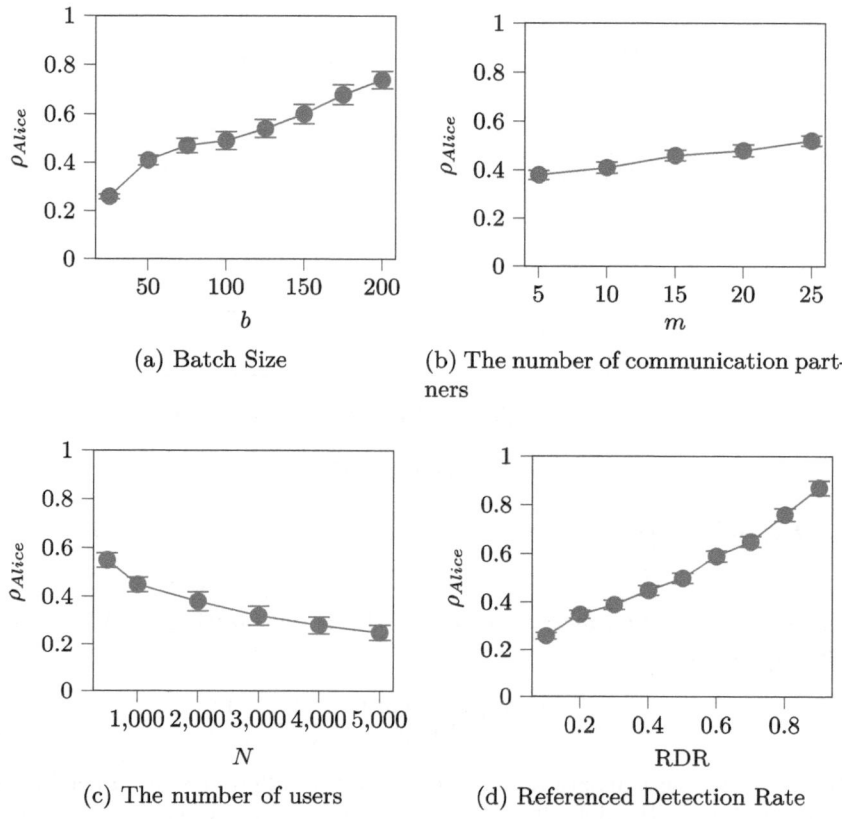

(a) Batch Size

(b) The number of communication partners

(c) The number of users

(d) Referenced Detection Rate

Fig. 5. The utilization values of Alice by varying system parameters and desired protection (RDR)

to identify partners [13]. Our results show that Alice's utilization increases with stronger protection provided by system parameters (see Fig. 5).

5   Discussion

We propose a management approach for users to self-control their anonymity protection. This model enables users to measure their current protection using feedback information from the anonymous communication system periodically. In our reference implementation, each user measures their current protection by applying the SDA to themselves using feedback information from the anonymous communication system. Then, they calculate the rate of correctly identified communication partners based on the current attack results as a measurement metric. If the calculated identification rate exceeds their desired level, they send dummy messages until their desired protection level is restored. Experimental studies show that following the proposed approach provides maintaining the desired level of protection during communication. Since only the users themselves know which of their contacts seems to be detected, and they would take protective actions that prevent consistent detection with high probability of these users, the adversary cannot be certain about the accuracy of the attack results. If Alice desires stronger protection, meaning she does not want any of her partners to be detected by the adversary, she can dynamically adjust the control unit to send dummy messages when the calculated values for her partners in the result vector $(\vec{v})$ approach the highest m. Because the adversary only consider the highest m values in the result vector. Therefore, users can take action before being de-anonymized.

The proposed model does not provide sufficient information for users to conduct attacks to de-anonymize others. The anonymity system only provides feedback on the recipients of the rounds. To apply a de-anonymization attack targeting any user, information about which rounds the targeted sender participates in is also necessary. Since each user only knows their own participation, they can only conduct attacks targeting themselves. If the recipients of the rounds are considered as information leakage, pseudonyms can be used to conceal the correct identity of recipients.

Experimental results show that the desired protection can be achieved by following the proposed approach. However, this approach may limit users regarding effective utilization of the anonymous communication network, depending on system parameters such as batch size. For example, in large user base systems, if the batch size is not properly set, excessive use of dummy messages may occur. This issue prevents users from sending their real messages to their communication partners. In this case, users have two options: either they may change their desired protection level to better utilize the system, or they can request system administrators to increase the batch size. In our future works, we will address the development of a management mechanism for setting proper system parameters dynamically based on users' requirements. We demonstrate the effectiveness of the proposed model in a basic Mix network. This model limits actions of users aimed at increasing anonymity protection to the use of dummy messages only. More complex models, such as the Stop-and-Go Mix [15], have adjustable parameters (e.g., average message delay, message-sending rate) that can be used to enhance anonymity protection by users. In this case, users have more option to increase the security level. Extending the proposed approach to these models would be an interesting direction for future studies.

6   Conclusions

In this paper, we propose a management model that distributes the task of protecting anonymity between users and the anonymous communication system. This model enables users to be aware of their current protection level and dynamically take their own actions to maintain their desired protection independently of the anonymous communication system. Our model does not assign static tasks to users, such as sending messages at exact times, nor does it impose significant overhead on anonymous communication systems to control the security of each user.

We demonstrate through experimental studies that the proposed approach maintains the desired level of anonymity protection during long-term communication. Additionally, we observed how system parameters affect the efficient utilization of users in the proposed model. For the experiments, we assume that the anonymous communication system operates as a basic mix model. In future studies, we would like to extend this approach to more complex mixing methods that provide stronger anonymity, thereby increasing user utilization while maintaining the desired level of anonymity. Additionally, we aim to develop models to set system parameters dynamically based on current user utilization as our future direction.

Acknowledgements. The first author would like to express gratitude to the Turkish Ministry of Education for funding his doctoral studies.

References

1. Agrawal, D., Kesdogan, D.: Measuring anonymity: the disclosure attack. IEEE Secur. Privacy **1**(06), 27–34 (2003). https://doi.org/10.1109/MSECP.2003.1253565
2. Aksoy, A., Kesdogan, D.: Effect of group based synchronization on user anonymity in mix networks. In: Proceedings of the 18th International Conference on Availability, Reliability and Security. ARES '23. Association for Computing Machinery, New York, NY, USA (2023). https://doi.org/10.1145/3600160.3604998
3. Cai, X., Zhang, X.C., Joshi, B., Johnson, R.: Touching from a distance: website fingerprinting attacks and defenses. In: Proceedings of the 2012 ACM Conference on Computer and Communications Security. CCS '12, pp. 605–616. Association for Computing Machinery, New York, NY, USA (2012). https://doi.org/10.1145/2382196.2382260
4. Chaum, D.L.: Untraceable electronic mail, return addresses, and digital pseudonyms. Commun. ACM **24**(2), 84–90 (1981)
5. Danezis, G., Dingledine, R., Mathewson, N.: Mixminion: design of a type iii anonymous remailer protocol. In: 2003 Symposium on Security and Privacy, pp. 2–15 (2003)
6. Danezis, G.: Statistical disclosure attacks. In: Gritzalis, D., De Capitani di Vimercati, S., Samarati, P., Katsikas, S. (eds.) SEC 2003. ITIFIP, vol. 122, pp. 421–426. Springer, Boston, MA (2003). https://doi.org/10.1007/978-0-387-35691-4_40
7. Diaz, C., Halpin, H., Kiayias, A.: The NYM network | the next generation of privacy infrastructure. https://nymtech.net/nym-whitepaper.pdf. Accessed 12 Apr 2023
8. Edman, M., Yener, B.: On anonymity in an electronic society: a survey of anonymous communication systems. ACM Comput. Surv. **42**(1) (2009)
9. Farber, D., Larson, K.: Network security via dynamic process renaming. In: Fourth Data Communications Symposium, pp. 8–13 (1975)
10. Gaballah, S., Nguyen, T.H.L., Abdullah, L., Zimmer, E., Mühlhäuser, M.: Mitigating intersection attacks in anonymous microblogging. In: Proceedings of the 18th International Conference on Availability, Reliability and Security. ARES '23, Association for Computing Machinery, New York, NY, USA (2023). https://doi.org/10.1145/3600160.3600166

11. Hayes, J., Troncoso, C., Danezis, G.: TASP: towards anonymity sets that persist. In: Proceedings of the 2016 ACM on Workshop on Privacy in the Electronic Society. WPES '16, pp. 177–180. Association for Computing Machinery, New York, NY, USA (2016). https://doi.org/10.1145/2994620.2994635

12. van den Hooff, J., Lazar, D., Zaharia, M., Zeldovich, N.: Vuvuzela: Scalable private messaging resistant to traffic analysis. In: Proceedings of the 25th Symposium on Operating Systems Principles. SOSP '15, pp. 137–152. Association for Computing Machinery, New York, NY, USA (2015). https://doi.org/10.1145/2815400.2815417

13. Kesdogan, D., Agrawal, D., Pham, V., Rautenbach, D.: Fundamental limits on the anonymity provided by the mix technique. In: 2006 IEEE Symposium on Security and Privacy (S P'06), pp. 14–99 (2006). https://doi.org/10.1109/SP.2006.17

14. Kesdogan, D., Buschkes, R.: Klassifizierung von Anonymisierungstechniken, pp. 321–332. Vieweg+Teubner Verlag, Wiesbaden (1999). https://doi.org/10.1007/978-3-322-89817-3_28

15. Kesdogan, D., Egner, J., Büschkes, R.: Stop- and- go-MIXes providing probabilistic anonymity in an open system. In: Aucsmith, D. (ed.) IH 1998. LNCS, vol. 1525, pp. 83–98. Springer, Heidelberg (1998). https://doi.org/10.1007/3-540-49380-8_7

16. Kesdogan, D., Pimenidis, L.: The hitting set attack on anonymity protocols. In: Fridrich, J. (ed.) IH 2004. LNCS, vol. 3200, pp. 326–339. Springer, Heidelberg (2004). https://doi.org/10.1007/978-3-540-30114-1_23

17. Kwon, A., Lu, D., Devadas, S.: XRD: scalable messaging system with cryptographic privacy. In: NSDI (2020)

18. Mallesh, N., Wright, M.: Countering statistical disclosure with receiver-bound cover traffic. In: Biskup, J., López, J. (eds.) ESORICS 2007. LNCS, vol. 4734, pp. 547–562. Springer, Heidelberg (2007). https://doi.org/10.1007/978-3-540-74835-9_36

19. Mathewson, N., Dingledine, R.: Practical traffic analysis: extending and resisting statistical disclosure. In: Martin, D., Serjantov, A. (eds.) PET 2004. LNCS, vol. 3424, pp. 17–34. Springer, Heidelberg (2005). https://doi.org/10.1007/11423409_2

20. Matloff, N.: Introduction to discrete-event simulation and the simpy language. Davis, CA. Dept of Computer Science. University of California at Davis, 1–33 (2008). Retrieved 2 August 2009

21. Parekh, S.: Prospects for remailers. First Monday 1(2) (1996).https://doi.org/10.5210/fm.v1i2.476

22. Piotrowska, A.M., Hayes, J., Elahi, T., Meiser, S., Danezis, G.: The loopix anonymity system. In: 26th USENIX Security Symposium (USENIX Security 17), pp. 1199–1216. USENIX Association, Vancouver, BC (2017)

23. Shen, T., et al.: DAEnet: making strong anonymity scale in a fully decentralized network. IEEE Trans. Depend. Secure Comput. 19(4), 2286–2303 (2022). https://doi.org/10.1109/TDSC.2021.3052831

24. Tyagi, N., Gilad, Y., Leung, D., Zaharia, M., Zeldovich, N.: Stadium: a distributed metadata-private messaging system. In: Proceedings of the 26th Symposium on Operating Systems Principles. SOSP '17, pp. 423–440. Association for Computing Machinery, New York, NY, USA (2017). https://doi.org/10.1145/3132747.3132783

25. Wolinsky, D.I., Syta, E., Ford, B.: Hang with your buddies to resist intersection attacks. In: Proceedings of the 2013 ACM SIGSAC Conference on Computer and Communications Security. CCS '13, pp. 1153–1166. Association for Computing Machinery, New York, NY, USA (2013). https://doi.org/10.1145/2508859.2516740

Machines Have Identities Too! Extending NIST's SP 800-63 for Device Identity

Tobias Hilbig$^{(\boxtimes)}$ [ID], Erwin Kupris [ID], and Thomas Schreck [ID]

HM Munich University of Applied Sciences, Munich, Germany
{tobias.hilbig,erwin.kupris,thomas.schreck}@hm.edu

Abstract. User authentication has evolved from simple password-based procedures to phishing-resistant biometric methods. NIST, in special publication 800-63, provides definitions and requirements for digital identities. However, there is a growing need to also identify and authenticate the device in use. Such information can be included in fine-grained policy decisions to further enhance an enterprise's security posture. In addition, device authentication has been described in the literature as a significant factor in zero trust architectures. Despite the adoption of this security architecture by major stakeholders, device authentication remains lacking. Therefore, we propose extensions to SP 800-63 that cover device identity aspects. In addition, we present a best-of-breed solution using FIDO2 and an extension for OpenID Connect. Our results demonstrate that the integration of device identity aspects is feasible and aligns well with the existing guidelines. The proposed scheme can pave the way for a future where device authentication will become the norm in enterprise networks.

Keywords: Digital Identity · Device Identity · Zero Trust Architecture · Identification · Authentication · Federation · OpenID Connect · FIDO2

1 Introduction

User authentication and authorization are well-researched topics. The "Digital Identity Guidelines" by NIST, published as SP 800-63 [1], specify definitions and requirements for the digital identity landscape of users. The publication encompasses the initial identity proofing and enrollment, authentication, and federation processes. For each activity in the identity lifecycle, NIST defines three assurance levels representing their strength and security level. High-assurance levels require the usage of two-factor authentication for user authentication.

It can be desirable to identify and authenticate the device in use in addition to the user. Such a network security approach is especially useful for high-security enterprise environments. In recent years, Zero Trust Architecture (ZTA) has become an important research area in this field. This network security paradigm can be summarized as "Never trust, always verify" [2]. In ZTAs, access to data

F. Martinelli and R. Rios (Eds.): STM 2024, LNCS 15235, pp. 36–46, 2025.
https://doi.org/10.1007/978-3-031-76371-7_3

and services must always be authenticated. In addition to the acting user, this must include device authentication. Device identity and state can be used to enhance policy-based access decisions.

Although device authentication has been discussed in the literature, its usage is relatively low in practice. In this work, we aim to bridge the existing, user-focused digital identity guidelines established by NIST with the emerging concept of device identity. The goal of this study is to evaluate the guidelines' applicability to devices and to propose suitable extensions. The proposed extensions can lay the foundation for future standardization efforts. Therefore, our study can pave the way toward a future where device authentication will become the default in enterprise networks, significantly enhancing security.

Our contributions are: (1) We first provide context by giving a detailed analysis of the NIST SP 800-63. (2) Building on this, we consider how each of the three sub-standards can be extended to cover devices. (4) We also propose an extension to OpenID Connect to include device assertions in federated environments. (3) Finally, we propose ways to realize device authentication in enterprise environments using modern authentication protocols such as FIDO2.

The remainder of this paper is organized as follows: In Sect. 2, we explore work related, while Sect. 3 provides the necessary background information. Extensions to the three sub-standards of the NIST SP are given in Sect. 4, Sect. 5, and Sect. 6, respectively. We discuss the proposed extensions and provide ideas for future work in Sect. 7. This paper concludes with Sect. 8.

2 Related Work

We surveyed the literature on device authentication: In an extensive literature review of wireless, physical layer authentication by Xie et al. [3], existing schemes were analyzed and categorized. A detailed comparison and evaluation of device authentication schemes in the Internet of Things (IoT) domain was conducted by El-hajj et al. in 2019 [4]. In their conclusions, the authors state that a combination of software and hardware authentication should be considered. Sehn et al. proposed BASA, a blockchain-based device authentication mechanism for industrial IoT [5]. Their solution allows the establishment of privacy-preserving trust across different domains. Designs for physically uncloneable functions (PUFs) were proposed by Suh et al. [6]. These designs enable the authentication of individual integrated circuits. They concluded that low-cost authentication and secure cryptographic key generation are achievable with PUFs. The examined literature mostly focused on device-to-device authentication without user involvement. In contrast, our study primarily examines scenarios that require user participation.

In addition to academic literature on device authentication, commercial offers that implement this concept exist. However, due to their proprietary nature, technical details are not publicly available. For example, Microsoft Entra [7] and Google BeyondCorp Enterprise [8] are enterprise security solutions that support device authentication features. While "Entra" is integrated into the Windows operating system, Google uses the Chrome browser to collect information about the device in use. Both offerings allow configuration of fine-grained access policies that take the device into account.

3 Background

We present existing methods for device authentication, discuss SP 800-63, and outline our considerations and adaptations of the guidelines regarding device identity.

3.1 Existing Device Authentication Methods

A broad range of authentication techniques exist that ensure authorized devices gain access to resources. Traditional methods include MAC and IP address filtering. They offer basic levels of control but are vulnerable to spoofing. Authentication via IEEE 802.1X [9] constitutes a more sophisticated approach. This includes basic authentication mechanisms based on preshared secrets and methods that leverage digital certificates and PKIs [10]. Based on a unique certificate issued to each device, TLS [11] with mutual authentication can be used at the application layer to validate the device identity. By combining certificates with hardware-based security features such as Trusted Platform Modules (TPMs), very high levels of assurance can be achieved on modern devices.

3.2 NIST Special Publication 800-63

The Digital Identity Guidelines, published by NIST as SP 800-63 [1], comprise a set of technical requirements that focus on digital identities. In the main document, NIST specifies their "Digital Identity Model". The sub-documents define assurance levels for each activity in the digital identity lifecycle. Finally, it offers guidance on digital identity risk management and structured processes for selecting appropriate levels for a given business case.

The digital identity model includes a high-level description of all entities involved in digital identity processes: A *subject*, typically a user, can play three roles depending on its state during transactions. It is called *applicant* when initiating the identity proofing process. After successful identity proofing, they are called *subscriber* and a *subscriber account* is created. The subscriber account is a collection of the subscriber's attributes and has one or more *authenticators*, i.e., *credentials*, bound to it. Subscribers who have completed the identity proofing process and want to authenticate themselves to a service are called *claimant*. After successful authentication, the user is recognized as a subscriber.

The so-called *Credential Service Provider (CSP)* identity-proofs applicants, manages the subscriber account, and binds the established authenticators to the subscribers. The *Verifier* handles claimants' authentication requests by verifying that they are in possession and control of the authenticators bound to the claimed subscriber account. In federated setups, *Identity Providers (IdPs)* commonly function as both the CSP and Verifier. They issue authentication assertions and relay them to *Relying Partys (RPs)*. RPs offer services to authorized subscribers. To make authorization decisions in a federated model, the RP depends on the information provided in the assertions generated by the IdP. In a non federated setup, the RP relies on the information contained in the subscriber account.

In the latest revision of the guidelines, some definitions have been adjusted to accommodate device identities in future versions. Furthermore, NIST states that device identity is not explicitly addressed; however, by referring to generic subjects throughout the document, the guidelines may also be applied to devices. However, no further adjustments regarding devices were made in the draft of the next revision. Here, we clarify some terms related to devices in a real-world enterprise context: For example, the CSP can be considered part of the asset management process a company employs for devices that interact with its services. If a device is identity-proofed, it is enrolled in the asset management. Similarly, a device that is identity-proofed but has no valid session is the claimant.

4   Enrollment and Identity Proofing

NIST SP 800-63A defines the requirements for the first activity related to digital identities, i.e., the identity proofing process and subsequent enrollment. First, we describe how NIST specifies user identities. We then propose our extensions to device identities and explain how we mapped them.

4.1   User Identities

The NIST SP defines three steps for the identity proofing flow: (1) The CSP collects identifying attributes from the applicant. (2) It validates these attributes in terms of authenticity, validity, and accuracy. (3) Finally, the CSP verifies the evidence, proves the applicant, and creates a subscriber account.

The three Identity Assurance Levels (IALs) represent the strengths of the identity proofing and enrollment processes. At IAL1, there is no requirement to correlate an applicant to a real-world identity, and all attributes collected during the process should be considered self-asserted. These attributes must be validated and verified for IAL2, creating a stronger correlation between the claimed identity and the real identity. To achieve IAL3, verification must be performed either in-person or remotely under supervision over a secure and trusted channel.

4.2   Device Identities

The process flow defined by NIST can generally be adopted for device enrollment. However, we must adjust the assurance levels to better portray the reality of enrolling a device into an enterprise's asset management. Our proposed levels, called Device Identity Assurance Levels (DevIALs), are built on top of each other.

DevIAL1: For the first level of device assurance, we propose to closely align with the definition by NIST regarding user identities. Similar to IAL1 for user identities, in the proposed DevIAL1, the CSP must not validate or verify attributes. While the CSP may request self-asserted attributes from the applicant device, this is not required. Therefore, any identifier and any additional attributes provided by the device are accepted without further validation or verification. As an example for DevIAL1, consider access to enterprise resources via a web browser. The user agent may serve as an identifying attribute which is not validated by the CSP. The resource being accessed sets a session cookie in the browser, which serves as the authenticator. Upon future visits, the resource can reliably determine that the browser in question accessed it in the past.

DevIAL2: For the second level, the identity proofing process must validate and verify the identity of the requesting device. The verification process involves checking whether the respective identifiers exist in device registries or other databases. Although such identifiers can be spoofed by other devices, the validation provides stronger security guarantees than the anonymous registration at DevIAL1. NIST defines several requirements for the second level, which we map to devices: The first requirement concerns Personally Identifiable Information (PII). This does not apply to devices because they do not provide PII. The second, third, and fourth requirements concern the strength of the evidence provided by the applicant. In this work, we refrain from categorizing all possible identifying attributes according to their strengths. For the second IAL, NIST allows for in-person and remote identity proofing, which we also allow for devices. In the device identity context, *in-person* means that the process is conducted or supported by a human, such as an administrator. *Remote* identity proofing is a fully automated process without human intervention. The remaining requirements, e.g., biometric data collection and trusted referees, are applicable only to humans; thus, they have no counterpart in device identities. An example for DevIAL2 at the application layer is the enrollment of a device into an asset management system based on the device serial number. The CSP checks whether this serial number is present in the order confirmation of the device manufacturer. If this is the case, the device is considered trustworthy and assigned an authenticator, e.g., an X.509 certificate.

DevIAL3: NIST tightens the validation and verification requirements for level three but does not add new requirements. In addition, remote identity proofing at this level is permitted only under certain restrictions: The entire process must be conducted under direct supervision of the CSP, e.g., via a high-resolution video link. As with IAL2, the requirements at this level are very specific to human users. Therefore, we restrict the allowed types of evidence to strong, cryptographically verifiable ones. Similar to people, devices can be issued verifiable identification attributes or credentials through authoritative and trusted parties, such as the device vendor. In addition, enterprises that enroll a device into their asset management system can validate these attributes to ensure the device's trustworthiness. For the third level, we propose that such credentials are protected by hardware measures to prevent malicious cloning. A vendor-assigned digital certificate stored in the device's TPM is an example for DevIAL3. The CSP can verify the certificate using the vendor's public key which provides a compelling assurance of the device's identity.

5   Authentication and Lifecycle Management

NIST SP 800-63B specifies user authentication and lifecycle management requirements. In the following, we describe NIST's approach to user authentication. We propose ideas to extend it to device authentication and define our so called Device Authentication Assurance Levels (DevAALs).

5.1   User Authentication

NIST specifies technical requirements and guidelines for secure user authentication processes. This includes the use of multi-factor authentication (MFA), password management, and cryptographic techniques. Three Authentication Assurance Levels (AALs) are defined to classify the strength and security of authentication processes: AAL1 represents the lowest level of assurance, requiring single-factor authentication, which may include a username and password or a single biometric factor. For AAL2, MFA is required to provide a higher degree of assurance. Two different factors, such as something you know, e.g., a password, and something you have, e.g., a mobile device, are required to enhance security. AAL3 offers the highest level of assurance, mandating the use of hardware-based authenticators with proof-of-possession and verifier impersonation resistance, and ensures the highest confidence in the identity assertion's validity. These levels allow organizations to select appropriate security levels based on information sensitivity and the potential risks associated with the authentication process.

5.2   Device Authentication

For human users, identity proofing and authentication procedures often differ; however, for devices, these processes are closely related. Unlike users, devices can only use technical properties, keys, and certificates for identity proofing. The same factors can also be used for authentication. Therefore, the proposed device authentication assurance levels align well with the DevIALs described in Sect. 4. Similar to NIST, we do not require DevIAL and DevAAL to match for individual devices. We differentiate the envisioned DevAALs along five characteristics, as shown in Table 1. In the following, we describe our proposed DevAALs and provide real-world examples.

DevAAL1: By definition, this level only provides *some* assurance about the authentication strength. For devices, this means that the CSP must not require a specific authentication method. The device is not required to protect the credential by specific means. This also allows credentials to be shared among multiple devices. Consider the following example for DevAAL1: At the network and

Table 1. Overview of Device Authentication Assurance Levels

Concept	DevAAL1	DevAAL2	DevAAL3
Authentication Method	any	CR	CR
Credential Protection	none	Software	Hardware
Credential Secrecy	no	yes	yes
Unique Credential	no	yes	yes
Spoofing possible	yes	yes	no

internet layers, MAC-based frame and IP-based package filtering can be applied. Session cookies can be considered at the application layer.

DevAAL2: We propose that at this level, the device must authenticate to the CSP via a cryptographically strong challenge-response (CR) protocol. Using such a protocol mitigates the risk of replay attacks and credential theft. The authentication method must also employ software measures to reduce the risk of disclosing credentials. Lastly, DevAAL2 uniquely identifies a device to the CSP. For example, authentication at the network layer can occur via 802.1X for DevAAL2. The client certificate used by the device can be stored without hardware protection, e.g., in the operating system certificate store.

DevAAL3: At the third level, hardware measures are required to protect credential data and ensure that they never leave the device. Similar to the second level, a cryptographically strong CR protocol is required. This enables a phishing-resistant authentication process that is protected by multiple layers of defense. For DevAAL3, 802.1X with the certificate protected by hardware measures, e.g., by a TPM, can be considered.

6 Federation and Future Approaches

In addition to analyzing NIST's considerations for federation, we propose an OpenID Connect (OIDC) extension and a comprehensive FIDO2-based solution.

6.1 Federation

NIST SP 800-63C specifies the requirements for federated operation and assertion transmission. The three Federation Assurance Levels (FALs) provided by NIST are directly applicable to devices. Therefore, no adjusted, device-focused FALs are required: While a bearer assertion signed by the issuer is required for FAL1, the assertion must also be encrypted for FAL2. To prevent malicious parties from using stolen assertions, signed and encrypted holder of key (HoK) assertions are required for FAL3. HoK assertions require that the party presenting the assertion to cryptographically proof the possession of the key referenced within the assertion itself.

6.2   Device Assertions via OpenID Connect

OIDC is a modern identity layer built on top of the OAuth 2.0 protocol [12], designed to facilitate user authentication in a standardized manner [13]. It allows federated RPs to verify the identity of an end user after authenticating at an authorization server, i.e., the IdP. OIDC employs a signed token, called ID token, that carries the user's identity information, thereby ensuring its verifiability. To send and receive device-specific assertions in federated setups, an extension to the OIDC protocol is required. The proposed extension covers device identities and allows both user and device authentication to occur simultaneously.

In OIDC, the RP can request user authentication using the *openid* scope. Similarly, RPs should be able to signal that device authentication should be performed by the IdP. To this end, we introduce the *device_auth* scope. If an IdP capable of device authentication receives an authorization request containing this scope, it should attempt to authenticate the device used in the transaction.

After successfully authenticating the device, the IdP generates a unique token we call the *device token*. This token is designed to carry device-specific attributes recorded in the device's subscriber account. In addition to these attributes, other relevant metadata about the device may be added to the device token, such as its DevAAL or dynamic information about the device's security posture. Like OIDC's ID token, the device token is signed by the IdP. This ensures that the token is secured against tampering, and the RP can trust that the device-specific information provided by the IdP is accurate and secure. A device token enhances security by allowing more granular access and control based on device-specific contexts, thereby providing a more robust framework for managing identities and access from multiple devices.

Finally, we define the new *deviceinfo* endpoint which functions similar to OIDC's established userinfo endpoint. This endpoint is specifically tailored to retrieve additional device-specific information, such as its hardware configuration, software versions, and security posture. The deviceinfo endpoint can be accessed by RPs possessing a valid access token, which allows them to query additional device attributes. This capability is particularly useful in environments where device integrity and context are critical for security measures and operational decision-making.

6.3   FIDO2 Solution

Following our definitions in Sect. 5, FIDO2 credentials [14] can be used to authenticate devices at various levels of assurance. Syncable passkeys and FIDO2 credentials stored on hardware security keys enable users to authenticate seamlessly using different devices. Therefore, these types of FIDO2 credentials cannot uniquely identify a device and thus only fulfill the requirements of DevAAL1. For higher assurance levels, only credentials bound to the exact device that is authenticating can be considered. Device-bound passkeys [15] represent such a FIDO2 credential, guaranteeing that the credential never leaves the device on which it was created. When these passkeys are stored in software and are not protected further, they satisfy the definition of DevAAL2. For DevAAL3, hardware protection of the credential is required. When a FIDO2 credential is stored within the TPM, it provides a robust layer of security by binding the user's identity to the device's hardware. In this case, the IdP can be sure that the credential is hardware-protected and can only be mapped to a single device; thus, the credential is highly resistant to phishing, theft, and replication attacks. Therefore, device-bound passkeys saved in the device's TPM can authenticate devices at DevAAL3. Thus, both the user and device can be authenticated simultaneously using FIDO2, providing a seamless and secure user experience. Although this best-of-breed procedure works in a web-based context in which a user is involved, it cannot be employed in machine-to-machine scenarios.

Table 2. Extended Digital Identity Assurance Levels Including Devices

Concept	Level	User Assurance	Device Assurance
Enrollment & Identity Proofing	IAL1	Self-asserted attributes	Anonymous
	IAL2	Remotely verified attributes	Weak identifier
	IAL3	In-person verified attributes	Strong identifier
Authentication & Lifecycle Management	AAL1	Single Factor	Any token
	AAL2	MFA	SW-based CR
	AAL3	HW-based MFA	HW-based CR
Federation & Assertions	FAL1	$sig\{$Bearer assertion$\}$	
	FAL2	$enc\{sig\{$Bearer assertion$\}\}$	
	FAL3	$enc\{sig\{$HoK assertion$\}\}$	

7 Discussion and Future Work

We present exemplary technologies for each assurance level of the NIST guidelines for user authentication and our device-focused extension in Table 2. In the following sections, we discuss our findings and provide directions for future work.

7.1 Discussion

We have proposed integrating of device assurance aspects into NIST SP 800-63. During our study, we noticed the following aspects that are worth highlighting: Similar to users, devices can have unique identifying attributes, such as serial numbers, MAC addresses, and IMEI numbers for mobile devices. For these attributes, we identified the following characteristics: (1) They can either be uniquely identifying or shared among a batch of devices. (2) Some are permanent, whereas others may change over time. (3) Attributes may be public, secret, or protected from being spoofed by hardware measures. In addition, devices do not necessarily need to enroll additional credentials after the identity proofing process. In fact, devices can reuse the identifying attributes used as a means to authenticate later. For example, a MAC address can be used as an identifying attribute during identity proofing and as a credential for network access. Consequently, identity proofing and authentication of devices are more aligned compared to these same processes regarding user identities. Furthermore, users typically authenticate at the application layer, unlike devices that can authenticate at various layers of the Internet model [16]. Finally, device authentication can either be closely related to user authentication or operate independently of user involvement. In the former case, both processes can occur simultaneously, e.g., when a user visits a website using a company-owned device. The latter case represents a scenario in which no human is involved, for example in machine-to-machine communication and IoT environments.

7.2   Future Work

We derive and develop directions for future work based on our findings: A viable extension to this work would be an analysis of all existing device enrollment and authentication protocols and procedures. These studies can assess these approaches along dimensions in terms of ease of use, technical complexity, and authentication strength. Furthermore, continuous access evaluation for user authentication is well established. How to continuously collect, transmit, and evaluate device-related information could be of interest. Moreover, we did not explicitly focus on workload identity aspects in our study. Future work should focus on how to consider workload-related information in machine-to-machine communication. Another direction for future work is the standardization of the proposed device assurance levels. Similarly, the proposed extensions to OpenID Connect need to be discussed by relevant standardization bodies. Similar to user identities, investigating the privacy implications of implementing device assurance levels, including data collection, storage, and sharing practices, is important. Future research could focus on mitigating potential privacy risks while maintaining a high level of assurance. Finally, the logical next step after device assurance is device trust, which was out of scope for this paper. Securely collecting relevant information from devices requires future attention.

8   Conclusion

Despite ongoing research and increasing industry interest, device authentication is not commonly performed. NIST, in its special publication 800-63, provides guidelines for digital identities with a focus on the user. In this paper, we propose ways to extend SP 800-63 to cover devices. Here, we discuss how the identity proofing, authentication, and federation aspects described by NIST can be mapped to devices. Our results demonstrate that extending the requirements to devices is indeed possible and aligns well with NIST's user-focused guidelines. We also presented an OpenID Connect extension for device assertions in federated environments. In addition, we have demonstrated that combining user and device authentication is possible using a highly assured, FIDO2-based solution. These proposals can lead to a future where device authentication will become the norm in enterprise contexts.

Competing interests. All authors declare that they have no conflicts of interest.

References

1. Grassi, P.A., Garcia, M.E., Fenton, J.L.: NIST Special Publication 800-63-3: Digital Identity guidelines. National Institute of Standards and Technology, Los Altos, CA (2017). https://doi.org/10.6028/NIST.SP.800-63-3
2. Buck, C., Olenberger, C., Schweizer, A., Völter, F., Eymann, T.: Never trust, always verify: a multivocal literature review on current knowledge and research gaps of zero-trust. Comput. Secur. **110**, 102436 (2021). https://doi.org/10.1016/j.cose.2021.102436
3. Xie, N., Li, Z., Tan, H.: A survey of physical-layer authentication in wireless communications. IEEE Commun. Surv. Tutor. **23**(1), 282–310 (2021). https://doi.org/10.1109/COMST.2020.3042188
4. El-Hajj, M., Fadlallah, A., Chamoun, M., Serhrouchni, A.: A survey of internet of things (IoT) authentication schemes. Sensors **19**(5), 1141 (2019)
5. Shen, M., et al.: Blockchain-assisted secure device authentication for cross-domain industrial IoT. IEEE J. Sel. Areas Commun. **38**(5), 942–954 (2020). https://doi.org/10.1109/JSAC.2020.2980916
6. Suh, G.E., Devadas, S.: Physical unclonable functions for device authentication and secret key generation. In: Proceedings of the 44th Annual Design Automation Conference. DAC '07, pp. 9–14. Association for Computing Machinery, San Diego, California (2007). https://doi.org/10.1145/1278480.1278484
7. Microsoft Corporation, Zero Trust Model - Modern Security Architecture — Microsoft Security, (2022) https://www.microsoft.com/en-us/security/business/zerotrust (visited on 08/19/2024)
8. Osborn, B., McWilliams, J., Beyer, B., Saltonstall, M.: Design to deployment at Google. Usenix Login **41**(1), 28–35 (2016)
9. IEEE Standard for Local and Metropolitan Area Networks–Port-Based Network Access Control. IEEE Std 802.1X-2020 (2020). https://doi.org/10.1109/IEEESTD.2020.9018454
10. Simon, D., Aboba, B., Hurst, R.: The EAP-TLS authentication protocol. Technical report, RFC 5216, Internet Engineering Task Force (2008)
11. Rescorla, E.: The Transport layer security (TLS) protocol version 1.3. Technical report 8446, 160 p. Internet Engineering Task Force (2018)
12. Hardt, D.: The OAuth 2.0 authorization framework. Technical report 6749, 76 p. Internet Engineering Task Force (2012)
13. Sakimura, N., Bradley, J., Jones, M., De Medeiros, B., Mortimore, C.: OpenID Connect Core 1.0, The OpenID Foundation (2014)
14. FIDO Alliance, FIDO2 - FIDO Alliance, (2024) https://fidoalliance.org/fido2/ (visited on 08/19/2024)
15. Hodges, J., Jones, J., Jones, M.B., Kumar, A., Lundberg, E.: Web authentication: an API for accessing public key credentials level 3. Technical report, W3C (2023)
16. Braden, R.T.: Requirements for internet hosts - communication layers. Technical report 1122, 116 p. Internet Engineering Task Force (1989)

Cryptography, Secure Computation
and Formal Methods

On β–Separating Sets and Deterministic Factoring

Jacek Pomykała[ID], Mariusz Jurkiewicz[(✉)][ID], Olgierd Onierczyk[ID],
and Kamila Prabucka[ID]

Faculty of Cybernetics, Military University of Technology, Warsaw, Poland
jacek.pomykala@wat.edu.pl, mariusz.jurkiewicz@wat.edu.pl,
olgierd.zolnierczyk@wat.edu.pl, kamila.prabucka@wat.edu.pl

Abstract. In this paper we consider the integer factorization problem that constitutes a base of security for cryptographic schemes from the family of solutions based on the Rivest-Shamir-Adleman concept. Apart from quantum Shor's algorithm, an efficient classical algorithm which enables to solve this problem in polynomial time has not been made so far. The best known classical algorithms carry out factorization in merely subexponential time. We show how a natural extension from the generalized approach to smoothness leads us to the concepts of decomposition witnesses, and on this basis, we present a new approach to elliptic-curve factorization. Instead of assuming that we have the witness of large order, what we did in [8], we assume there is given the set X of witnesses generating large subgroup in $G \subset E(\mathbb{Z}_N)$ that have some additional properties. We justify that either X contains the separating witness or the subgroup $< X >$ generated by the set X in $E(\mathbb{Z}_N)$ is noncyclic. The main result of this paper shows how to factor an integer, which is a product of two primes, in time $L\left(\frac{1-\theta_\sigma(1-\beta)}{\sigma\alpha_0} + \sigma\alpha_0 + o(1),\ r\right)$. This is an improvement as compared to the previous results provided that $\theta_\sigma \geq (2 - 2\sigma + \sigma^2)/2(1 - \beta)$.

Keywords: Elliptic curve · Frobenius traces modulo primes · Integer factorization

1 Introduction

The integer factorization problem is to find prime factors decomposition of a given positive integer N. Even though by the fundamental theorem of arithmetic the solution to this problem always exists, an efficient algorithm which enables to conduct this task in polynomial time has not been made so far. The best known classical algorithm GNFS does factorization in merely subexponential time. In fact, there are two main brunches that constitute pillars for integer factorization methods and apart from sieving, whose representative is GNFS, there is also a popular approach based on smoothness. Smooth numbers are numbers with only small prime factors. They have both a simple multiplicative structure and are

© The Author(s), under exclusive license to Springer Nature Switzerland AG 2025
F. Martinelli and R. Rios (Eds.): STM 2024, LNCS 15235, pp. 49–65, 2025.
https://doi.org/10.1007/978-3-031-76371-7_4

also fairly numerous. Due to these properties, smooth numbers play a key role in integer factorization algorithms like a famous Pollard $p-1$ method and its follower ECM [6] which is based on elliptic curves over $\mathbb{Z}_N$, for instance.

In [9] the definition of admissible elliptic curves and separating together with non-separating witnesses Q for N were defined and the related β–Conjecture concerning the admissible curves was stated. The result concerning the average case time complexity for separating witnesses was given in [8]. In somewhat complementary case of non-separating witness the benefit for the factoring of N comes from the value of the parameter $\gamma > 1/3$, which implies that parameter β also should be greater than $1/3$ In this approach we deal with the case of arbitrary value $\beta > 0$ with the aim of proving somewhat different result concerning the decomposition of $N = pq$. Namely, instead of assuming that we have the witness Q of large order greater than $r^{1-\gamma}$ we assume that we have the set X of witnesses generating large subgroup in $G \subset E(\mathbb{Z}_N)$ of local order of $E(\mathbb{F}_r)$ greater or equal to $r^{1-\eta}$ (assuming we have the isomorphism $E(\mathbb{Z}_N) \approx E(\mathbb{F}_p) \times E(F_q)$). We will prove that (under the sutable assumptions) in such case we either prove that X contain the separating witness Q or prove that the subgroup $< X >$ generated by the set X in $E(\mathbb{Z}_N)$ is noncyclic.

2   Preliminaries on Pseudo-Elliptic Curves

In this section we recall basic facts on elliptic curves over $\mathbb{Z}_N$, where $N = \prod_{i=1}^{s} p_i$ is coprime to 6 (see [5,6]). The projective plane $\mathbb{P}^2(\mathbb{Z}_N)$ is defined to be the set of equivalence classes of primitive triples in $\mathbb{Z}_N^3$ (i.e., triples (x_1, x_2, x_3) with $\gcd(x_1, x_2, x_3, N) = 1$) with respect to the equivalence $(x_1, x_2, x_3) \sim (y_1, y_2, y_3)$ iff $(x_1, x_2, x_3) = u(y_1, y_2, y_3)$ for a unit $u \in \mathbb{Z}_N^*$. An elliptic curve over $\mathbb{Z}_N$ is given by the short Weierstrass equation $E : y^2 z = x^3 + axz^2 + bz^3$, where $a, b \in \mathbb{Z}_N$ and the discriminant $-16(4a^3 + 27b^2) \in \mathbb{Z}_N^*$. The point $O = (0 : 1 : 0)$, called the zero point, belongs to $E(\mathbb{Z}_N)$. Let $V(E(\mathbb{Z}_N)) = \{(x, y) \in E(\mathbb{Z}_N)\} \cup \{O\}$ be the set of finite points in $E(\mathbb{Z}_N)$ with the zero point O. For each point $(x : y : z) \in E(\mathbb{Z}_N) \setminus V(E(\mathbb{Z}_N))$ the $\gcd(z, N)$ is a nontrivial divisor of n.

Let $E(\mathbb{F}_{p_i})$ be the group of $\mathbb{F}_{p_i}$-rational points on the reduction $E \mod p_i$ for primes $p_i \mid N$. If $E(\mathbb{Z}_N)$ is the set of points in $\mathbb{P}^2(\mathbb{Z}_N)$ satisfying the equation of E then by Chinese Remainder Theorem there exists the bijection

$$\varphi : E(\mathbb{Z}_N) \to \prod_{i=1}^{s} E(\mathbb{F}_{p_i}) \tag{2.1}$$

induced by the reductions $\mod p_i$. The points $(x : y : z) \in E(\mathbb{Z}_N)$, with $z \in \mathbb{Z}_N^*$, can be written as $(x/z : y/z : 1)$ and are called finite points. The set $E(\mathbb{Z}_N)$ is a group with the addition for which φ is a group isomorphism, which in general can be defined using the so-called complete set of addition laws on E (see [5]).

To add two finite points $P, Q \in E(\mathbb{Z}_N)$ we can also use the same formulas as for elliptic curves over fields in the following case: for $\varphi(P) = (P_1, \ldots, P_s)$ and

$\varphi(Q) = (Q_1, \ldots, Q_s) \in \prod_i E(\mathbb{F}_{p_i})$ either $Q_i \neq \pm P_i$ for each i or $Q_i = P_i$ and $Q_i \neq -P_i$ for each i. Then

$$\begin{cases} x_{P+Q} = \lambda^2 - x_P - x_Q \\ y_{P+Q} = \lambda(x_P - x_{P+Q}) - y_P, \end{cases} \qquad (2.2)$$

where

$$\lambda = \begin{cases} \frac{y_Q - y_P}{x_Q - x_P} & \text{if } Q_i \neq \pm P_i \text{ for each } i \\ \frac{3x_P^2 + a}{2y_P} & \text{if } Q_i = P_i \text{ and } Q_i \neq -P_i \text{ for each } i. \end{cases}$$

Let $P, Q \in E(\mathbb{Z}_N)$ and if $R = P + Q$ is finite then the formulas (2.2) give the coordinates of the resulted point R. Otherwise, either we find a nontrivial divisor of N or prove that all local orders $\operatorname{ord} R_i$ are equal to each other for $i = 1, 2, \ldots, s$ (see e.g. [3,6] for details).

In what follows we assume that $N = pq$ has two distinct prime divisors (both > 3) and $B = B(N)$ is fixed. We apply the above formulas to compute the point $mQ \in E(\mathbb{Z}_N)$. The computation of finite point mQ takes $O(\log m)$ adding operations in $E(\mathbb{Z}_N)$. For B-smooth number $m = m_B$ represented as $m = p_k^{e_k} \ldots 3^{e_3} 2^{e_2}$, where $e_i = e_i(m)$ is the highest exponent in which p_i does not exceed $\min(p, q) + 2\sqrt{\min(p, q)} + 1$ the computation of $m_B Q$ takes

$$\ll \log N \sum_{i \leq k} \log(p_i) = O(B \log N) \qquad (2.3)$$

adding operations in $E(\mathbb{Z}_N)$ in view of the Prime Number Theorem. The next result comes from [4] (see Proposition 2.5)

Proposition 1. *Let E be an elliptic curve over $\mathbb{Z}_N$ and $Q \in E(\mathbb{Z}_N)$ be a finite point such that $m_D Q_p = O$ for some prime $p \mid N$ and m_D being the product of all primes $r \leq D$ in suitable powers ν_r such that $r^{\nu_r} < 2N$. Assume that*

$$\operatorname{ord}(Q_p) \neq \operatorname{ord}(Q_q) \qquad (2.4)$$

for some prime $q \mid N$. Then one can obtain a nontrivial divisor of N either computing $m_D Q$ successively as $r_k^{\nu_k} \cdots 3^{\nu_3} 2^{\nu_2} Q \in E(\mathbb{Z}_N)$, or doing some additional computations which (in the worst case) take $O(D^2 \log D \log N) = D^{2+o(1)}$ bit operations in $E(\mathbb{Z}_N)$, as N tends to infinity.

More accurate analysis of partially B-smooth orders of $E(\mathbb{F}_r)$ allows to accelerate the complexity to $B^2 + D^{\sigma + o(1)}$ ($\sigma < 2$) of the related algorithm, compared to the deterministic time $D^{2+o(1)}$ above. Namely referring to definitions of admissible curves and decomposition witnesses stated in Sect. 3 we have the following extension of the above result

Proposition 2. *(see [9]) Let N ($N = pq$, $p < q < \vartheta p$) be given, $\{r, r'\} = \{p, q\}$ and E be an $(I_r, B, \beta, D, \sigma; 0)$-admissible curve (see definition in the next Sect. 3). Assume that the point $Q = (x, y) \in E(\mathbb{Z}_N)$ is E-strong or E-weak separating witness for N. Then one can find p and q in deterministic time $O\left((B^2 + D^\sigma)^{1+o(1)}\right)$.*

3   Admissible Elliptic Curves

We recall here the definition of admissible elliptic curves over $\mathbb{Z}_N$ and then the related witnesses for N (cf. [9]).

We consider $E = E_{\bar{b}}$ over $\mathbb{Z}_N$ with $\bar{b} = (b_1, b_2)$ given by the short Weierstrass equation $y^2 = x^3 + b_1 x + b_2$ (where $\Delta_{\bar{b}} := 6(4b_1^3 + 27b_2^2)$ is coprime to N).

Definition 1. *The number $m \in \mathbb{N}$ is called (r, B, β, D, σ)−admissible if it is (B, β, D)−smooth, $m \in I_r$ and the following conditions hold*

$$\beta \leq 1 - \frac{\sigma \log D}{\log I_r^-}, \tag{3.1}$$

$$\left(\frac{m}{s_B(m)}\right)^* \leq D^\sigma. \tag{3.2}$$

Definition 2. *We say that E over $\mathbb{Z}_N$ is $(B, \beta, D, \sigma; \gamma)$-admissible if E_r is (I_r, B, β, D)-admissible number for some $r \in \{p, q\}$ and additionally it holds*

$$1 \leq \min(a_p(E), a_q(E)) \leq 2p^{1/2-\gamma}. \tag{3.3}$$

Given N, $B = B(B), D = D(N)$ we say that the curve E over $\mathbb{Z}_N$ is (B, β, D, σ) admissible if E_r is (I_r, B, β, D)−admissible for some $r \in \{p, q\}$. We will shortly say that E is (β, σ, γ)-admissible (or shortly admissible curve), if the values of the remaining parameters are clear from the context. In what follows we let $B < D$, $\beta \in [0, 1], \sigma \in [1, 2]$ and $\gamma \in [0, 1/2]$ andconsider the interval $I := I_r = [I_r^-, I_r^+]$, where $I_r = [r + 1 - \sqrt{r}, r + 1 + \sqrt{r}]$, where $r \in \{p, q\}$.

4   Separating Witnesses and Main Results

4.1   Introductory Remarks

In this paper we focus on the deterministic approach based on the elliptic curve factoring. The principal idea is to consider the pair $(E, \bar{X})$, where $\bar{X}$ is the sets of points of the elliptic curve E over the ring $\mathbb{Z}_N$, for semiprime number $N = pq$. We prove the results on the existence in the set $\bar{X}$ the points $\bar{P}$ which allow to factor N in a given deterministic time $t = t_A$. The results depend on the cyclicity of the group $G_{\bar{X}}$ and the multiplicative structure of the local orders E_r of elliptic curve $E(\mathbb{Z}_N)$ depending on the smoothness levels B and $D > B$ of local orders E_r.

The special case, when B and D are subexponential functions, was the basic challenge in papers [9] and [8] where some computational support on the related β−Conjecture was presented. The main contribution concerned the proper definition of admissible curve E, and $(E, B)-$, (E, D)−separating witnesses on the one hand and the nonseparating witnesses for N on the other. This approach allows to obtain the (conditional) results on deterministic decomposition $N = pq$

depending on the parameters $B < D$ and additional parameters $\beta \in [0,1]$ and $\sigma \in [1,2]$.

In this approach we apply two approaches concerning the existence of the related (E, B)–separating witnesses and (E, D)–separating witnesses based on the application of the Silver-Pohlig-Hellman [7] algorithm applied for the factorization problem in [10], extended for the elliptic-curve approach.

4.2   Basic Notions and Notations

Let $p < q < \vartheta p$ and $d \in \{p, q, N\}$. By $G_{\bar{X}}$ we denote the subgroup in $E(\mathbb{Z}_N)$ generated by the set $\bar{X} \subset E(\mathbb{Z}_N)$. Let $\eta \in [0, 1/2)$ and $B = B(N)$, where $N \geq N_0(\eta)$ is sufficiently large.

Let $\pi_r(Q)$ be the projection/reduction of (admissible point $Q \in E(\mathbb{Q})$) on $E(\mathbb{Z}_r)$ where $r \in \{N, p, q\}$. By Φ_N^B we mean the homomorphism sending any $Q = (x, y)$ represented by CRT as (Q_p, Q_q) to the multiples points pair $\left(\frac{E_p}{s_B(E_p)} Q_p, \frac{E_q}{s_B(E_q)} Q_q \right) \in E(\mathbb{F}_p) \times E(\mathbb{F}_q)$.

The main results concern (E, B) and (E, D)–separating witnesses, where E is (B, β)– smooth and (B, β, D, σ)–admissible, respectively. Let $\bar{X} \subseteq E(\mathbb{Z}_N)$ be a nonempty set. We will prove the results on the existence of separating witnesses for N in terms of the structure of the groups $G_{\bar{X}}$ and its homomorphic image $G_{B, \bar{X}}$ under Φ_N^B and its consequences for deterministic factoring of N.

4.3   Conditional Results on Factoring N Based on the Knowledge of B–Factor of E_r

In this section we formulate and prove the first result concerning the partial knowledge about E_r. Let $p < q < \vartheta p$. Let $B_0 := \min_{r \in \{p, q\}} r + 1 + 2\sqrt{r}$, $\{r, r'\} = \{p, q\}$.

Lemma 1. *(Coppersmith) (see [1]) If we know $N = pq$ and the high order $(1/4)(\log_2 N) + O_\vartheta(1)$ bits of q, where $p < q < \vartheta p$, then in polynomial time in $\log N$ we can discover p and q.*

Directly from the above lemma we deduce that if E_r is known for some $r \in \{p, q\}$ then one can discover the decomposition $N = pq$ in deterministic polynomial time. Let us remark here that in the decomposition of E_N is known and $\omega(E_M) = O(\log \log N)$, then the deterministic polynomial time factorization of $N = pq$ follows from [2].

For any point $\bar{R} \in E(\mathbb{Z}_N)$ we denote by R and R' the reduced points in $E(\mathbb{Z}_r)$ and $E(\mathbb{Z}_{r'})$ respectively. By $G_{\bar{X}}$ we mean the subgroup of $E(\mathbb{Z}_N)$ generated by the set $\bar{X}$. We have

Theorem 4.1. *Let $\gamma \in (0, 1/2]$. The decomposition $N = pq$ can be computed in deterministic polynomial time polylogN (depending on c, c', ϑ) if at least one of the following conditions is satisfied*

(i) *If $s_{B_0}(E_r)$ is known*

(ii) *If $d := \gcd\left(s_B(E_r)s_B(E_{r'})\right) \geq c(\vartheta)r^{1-\gamma}$ is known and either for some positive constant $c > 0$ it holds*

$$1 \leq \min(a_p(E), a_q(E)) \leq r^{1/2-\gamma} \qquad (4.1)$$

or

$$d \geq r^{1-c\frac{\log\log N}{\log N}} \qquad (4.2)$$

(iii) *If we know the point $\bar{R} \in G_{\bar{X}}$ and $d' \geq c'(\vartheta)r^{1-\gamma}$ such that $d' = s_B(\mathrm{ord}R) = s_B(\mathrm{ord}R')$ and for some positive constant $c > 0$ it holds that either*

$$1 \leq \min(a_p(E), a_q(E)) \leq r^{1/2-\gamma} \qquad (4.3)$$

or

$$d' \geq r^{1-c'\frac{\log\log N}{\log N}}. \qquad (4.4)$$

Proof. The first conclusion follows from Lemma 1 above. The second assertion follows by the application of Lemma 1 and Theorem [9]. The third claim follows from the assertion (ii), since $\gcd\left(s_B(E_r), s_B(E_{r'})\right) \geq \max\left(s_B(\mathrm{ord}R), s_B(\mathrm{ord}R')\right)$.

4.4 $(E, B)-$Separating Witnesses

In this section we state and prove the results on the sufficient condition for the set $\bar{X}$ to contain the $(E, B)-$separating witness (see definition below) and, as a consequence the decomposition $N = pq$ in deterministic time $|\bar{X}|B^{2+o(1)}$. We start from recalling that $G_{\bar{X}}$ stands for the subgroup of $E(\mathbb{Z}_N)$ generated by the set $\bar{X} \in E(\mathbb{Z}_N)$ and $G_{B,\bar{X}}$ being its homomorphic image under the homomorphism Φ_N^B.

Definition 3. *Let E_r be $B-$smooth number for some $r \in \{p, q\}$. The point $\bar{P} \in E(\mathbb{Z}_N)$ is called $(E, B)-$separating witness if*

$$s_B(\mathrm{ord}P) \neq s_B(\mathrm{ord}P'), \qquad (4.5)$$

where $P = \pi_r(\bar{P})$ and $P' = \pi_{r'}(\bar{P})$ and $\{r, r'\} = \{p, q\}$.

Theorem 4.2. *Let $N = pq$, where $p < q < \vartheta p$, $E(\mathbb{Z}_r)$ is cyclic group for each $r \in \{p, q\}$. Assume that for a set $\bar{X}$, $0 \leq \eta \leq 1/2$ and for some (explicitly determined) constant $c = c(\vartheta)$ we have that $G_{B,\bar{X}}$ is cyclic and we have*

$$|G_{\bar{X}}| \geq N^{1-\eta} \qquad (4.6)$$

$$s_B(E_r) \geq c(\vartheta)r^{2\eta}. \qquad (4.7)$$

Then the set $\bar{X}$ contains some $(E, B)-$separating witness and in deterministic time $|\bar{X}|B^{2+o(1)}$ we can discover decomposition $N = pq$.

Proof. We will prove that $\bar{X}$ contains an element $\bar{P} = (P, P')$ such that

$$s_B(\text{ord}P) \neq s_B(\text{ord}P'), \tag{4.8}$$

provided the subgroup $G_{B,\bar{X}}$ is cyclic.

In this connection consider the image $\Phi_N^B(\pi_N(\bar{X}))$ of homomorphism

$$\Phi_N^B : E(\mathbb{Z}_N) \to E(\mathbb{Z}_N).$$

We have the equality

$$|G_{B,\bar{X}}| = |G_{\bar{X}}| / \#ker\Phi_N^B \tag{4.9}$$

Assume on the contrary that the conditions 4.10 and 4.11 hold, but the condition 4.8 does not hold. Let $\bar{R} = \left(\frac{E_p}{s_B(E_p)} P, \frac{E_q}{s_B(E_q)} P' \right) := (R, R')$. The order of point $\bar{R}$ is a divisor of the lcm of the related local orders $\text{ord}R$ and $\text{ord}R'$ respectively, hence we have

$$\text{ord}\left(\Phi_N^B(\pi_N(\bar{P}))\right) = \text{ord}\left(\Phi_N^B(P, P')\right) \mid \text{lcm}\left(\text{ord}\Phi_N^B(P), \text{ord}\Phi_N^B(P')\right)$$

$$= \text{lcm}\left(\text{ord}\frac{E_p}{s_B(E_p)} P, \text{ord}\frac{E_q}{s_B(E_q)} P'\right)$$

hence it divides

$$\text{lcm}\left(\text{ord}R, \text{ord}R'\right) \mid \text{lcm}\left(s_B(\text{ord}P), s_B(\text{ord}P')\right) = s_B(\text{ord P})$$

for all $\bar{P} = (P, P') \in G_{\bar{X}}$. Since $G_{B,\bar{X}}$ is cyclic we obtain that the order of the group $G_{B,\bar{X}}$ is equal to

$$\text{lcm}_{\bar{P} \in \bar{X}} \text{ord}\Phi_N^B(\bar{P}) = \text{lcm}_{\bar{P} \in \bar{X}} s_B(E_p) = s_B(E_p).$$

Now we give a lower bound for the numerator and upper bound for the denominator in 4.9 above to obtain the contradiction.

Namely the numerator is $\geq N^{1-\eta}$ by (4.10). To give an upper bound for the denominator we remark that in $ker\Phi_N^B$ we count the points $\bar{P} = (P, P'))$ such that both $\frac{E_p}{s_B(E_p)} P$ and $\frac{E_q}{s_B(E_q)} P'$ are infinite points on the curves $E(\mathbb{Z}_r)$, $r \in \{p, q\}$, hence by cyclicity of $E(\mathbb{Z}_r)$ their number is by CRT equal to

$$\# ker\, \Phi_N^B = \phi\left(\frac{E_p}{s_B(E_p)}\right)\phi\left(\frac{E_q}{s_B(E_q)}\right) \leq \frac{E_p E_q}{s_B(E_p)s_B(E_q)}$$

$$< \frac{2N}{s_B(E_p)s_B(E_q)}$$

for sufficiently large $N \geq N_0$. Summing up we obtain that

$$\frac{N^{1-\eta}}{2N/s_B(E_p)s_B(E_q)} < |\Phi_N^B(G_{\bar{X}})| \leq s_B(E_p)$$

giving the contradiction since $s_B(E_r) > c(\vartheta)r^{2\eta}$. The second assertion follows by Proposition 1 by checking all points $\bar{P} \in \bar{X}$. This completes the proof of Theorem 4.2.

Directly from Theorem 4.1 and Theorem 4.2 we deduce the following corollary.

Corollary 1. *If $G_{\bar{X}}$ is cyclic and E_r is coprime to $r-1$ for $r \in \{p,q\}$ then one can factor N in deterministic time $|\bar{X}|B^{2+o(1)}$, provided*

$$|G_{\bar{X}}| \geq N^{1-\eta} \tag{4.10}$$

$$s_B(E_r) \geq c(\vartheta)r^{2\eta}. \tag{4.11}$$

Assume that we know $d' \geq c(\vartheta)r^{1-\gamma}$ such that either $0 \leq \gamma = \gamma(N) \leq c' \frac{\log\log N}{\log N}$ or

$$1 \leq \min(a_p(E), a_q(E)) \leq r^{1/2-\gamma}. \tag{4.12}$$

and for some $\bar{P} \in \bar{X}$ we have that

$$s_B(\text{ord}P) = s_B(\text{ord}P') = d'. \tag{4.13}$$

Then we can factor N in deterministic time $|\bar{X}|polylogN$.

Proof. The first assertion follows from Theorem 4.1 since the condition that $G_{\bar{X}}$ is cyclic implies that $\Phi_N^B(G_{\bar{X}})$ is also cyclic. The second follows since we check all the $|\bar{X}|$ points $\bar{P}$ and if at least one satisfies the condition 4.13 then we decompose $N = pq$ in deterministic time $|\bar{X}|polylogN$.

Remark. By the group structure of $E(\mathbb{Z}_r)$ we have that if E_r is coprime to $r-1$ then the group $E(\mathbb{Z}_r)$ is cyclic. Moreover one can easily deduce that the condition that E_r is coprime to $r-1$ follows from two conditions $a_r(E) \neq 2$ and $P_-(r-1) > |a_r(E) - 2|$.

In the following section we consider the somewhat complementary case when $G_{\bar{X}}$ is not cyclic group.

4.5 (E, D)−Separating Witnesses

Theorem 4.2 gives the sufficient condition to factor N in deterministic time $|\bar{X}|B^{2+o(1)}$ which contains the assumption that the image $\Phi_N^B(G_{\bar{X}})$ is cyclic. If it is not cyclic then certainly $G_{\bar{X}}$ also is noncyclic. In the next section we consider such case assuming that the local order E_r is D−smooth for a given $D > B$ provided E satisfies the related admissibility condition. It is based on the algorithm $\mathcal{A}$ described in the next section. We begin with thew relevant definitions.

Definition 4. Let E be (B, β, D, σ)−admissible number. The point $\bar{P}$ is called (E, D)−separating witness if

$$s_D(\text{ord}P) = \text{ord}P \neq \text{ord}P', \tag{4.14}$$

where $P = \pi_r(\bar{P})$ and $P' = \pi_{r'}(\bar{P})$.

Definition 5. *Let E_r be (B, β, D, σ)−admissible number. The set $\bar{X}$ is called $t_{\mathcal{A}}$−separating set if the algorithm $\mathcal{A}$ finds (E, D)−separating witness $\bar{P} \in \bar{X}$ or declares that such $\bar{P}$ does not exist in $\bar{X}$, in deterministic time $t_{\mathcal{A}} = t_{\mathcal{A}}(B, D, \bar{X})$.*

In Corollary 1 the conditions for elliptic curve E were given so that the set $\bar{X}$ was $t_{\mathcal{A}} = |\bar{X}| B^{2+(1)}$ separating set when $G_{\bar{X}}$ was cyclic group. Here we will prove the analogous result depending on additional parameters D, β, σ proving that the set $\bar{X}$ is $t_{\mathcal{A}}$−separating with $t_{\mathcal{A}} = t_{\mathcal{A}}(B, D, \bar{X}) = |\bar{X}| \left(B^2 + D^{3/2}\right)^{1+o(1)}$ when $G_{\bar{X}}$ is not cyclic group. Namely we will prove the following theorem.

Theorem 4.3. *Let $N = pq$, where $p < q < \vartheta p$, E is (B, β, D, σ)−admissible, $\bar{X} \subseteq E(\mathbb{Z}_N)$, $\sigma \le 3/2$, $0 \le \eta \le \min(1/2, \beta)$ and $c = c(\vartheta)$ be the related (explicitly determined), positive constant.*

Assume that $E(\mathbb{Z}_r)$ is cyclic for each $r \in \{p, q\}$ but $G_{\bar{X}}$ is not cyclic. Then the algorithm $\mathcal{A}$ given below finds (E, D)−separating witnesses $\bar{P} \in \bar{X}$ and $\bar{Q} \in \bar{X}$ such that the group $G_{\{\bar{P}, \bar{Q}\}}$ is not cyclic and discovers decomposition $N = pq$ in deterministic time $t_{\mathcal{A}} = B^{2+o(1)} + |\bar{X}| D^{3/2+o(1)}$, hence $\bar{X}$ is $t_{\mathcal{A}}$−separating set.

Analogously as in Corollary 1 if the set $\bar{X}$ contains the point $\bar{P}$ such that we know the value $d \ge c(\vartheta) r^{1-\gamma}$ such that either $0 \le \gamma \le c' \frac{\log \log N}{\log N}$ or

$$1 \le \min(a_p(E), a_q(E)) \le r^{1/2-\gamma} \tag{4.15}$$

and for some $\bar{P} \in \bar{X}$ we have that

$$s_D(\mathrm{ord} P) = \mathrm{ord} P = \mathrm{ord} P' = d,$$

then the decomposition $N = pq$ can be discovered in deterministic time $|\bar{X}| \mathrm{polylog} N$.

Remark. If $|G_{\bar{X}}| > N^{1-\eta}$ and $G_{\bar{X}}$ is cyclic then we can reduce the search for the element of large order in $G_{\bar{X}}$ to analysing subsequently the orders of groups $G_{\{\bar{P}, \bar{Q}\}}$ and if we find $\bar{R} \in G_{\{\bar{P}, \bar{Q}\}}$ generating it then we will find the element of large order in $G_{\bar{X}}$ and both the conditions $s_B(E_r) \ge r^{2\eta}$ and $s_B(\mathrm{ord} \bar{P}) \ge r^{1-\eta}$ are quite realistic (unless B is too small). In such case, having better than the conventional way of searching the candidates for possible values of E_r one can prove that in $\bar{X}$ the (E, D)−separating witness exists unless all the local orders have the same and large D−factors. In the last case we can apply the theorem below.

Theorem 4.4. *Let E be (B, β, D, σ)−admissible elliptic curve, $\bar{X} \subseteq E(\mathbb{Z}_N)$, such that*

$$0 < \eta(N) < \min(1/4, \beta(N))$$

If the algorithm $\mathcal{A}$ fails in finding the (E, D)−separating witness for N then one can factor N nontrivially in deterministic time $|\bar{X}| \mathrm{polylog} N$ at least in one of the following cases

(1) *Assume that* $|G_{\bar{X}}| > N^{1-\eta}$, *E is admissible and*

$$1 \le \min(a_p(E), a_q(E)) \le 2p^{1/2-\eta} \qquad (4.16)$$

(2) $\eta(N) = O(\frac{\log\log N}{\log N})$

Proof. To deduce the first assertion we remark that if the algorithm $\mathcal{A}$ does not return $(E, D)-$separating witness $\bar{P}$ then it returns the point $\bar{P}$ generating $G_{\bar{X}}$, hence the reduced point P has the order $\ge cr^{1-\eta}$ for some $\eta < 1/4$ and the condition 4.16 implies the result in view of [9]. The second assertion follows from Lemma 1.

Theorem 4.5. *If in the set* $\bar{X}$ *there is no* $(E, D)-$ *separating witness then algorithm* $\mathcal{A}$ *returns the generator of* $G_{\bar{X}}$ *of group order* $G_{\bar{X}}$.

5   Description of Algorithm $\mathcal{A}$ and Examples

5.1   Algorithm

The algorithm below is based on the Silver-Pohlig-Hellman algorithm applied by raek [10] for the ring $\mathbb{Z}_N$. In what follows we assume that $E = E_{\bar{b}}$ and the finite points $(x, y) \in E(\mathbb{Z}_N)$ have integral coordinates. We start from the initiary subroutine of the the algorithm in the case when $\bar{X} = \{\bar{P}, \bar{Q}\}$. Let $l \in (B, D]$ be a prime number and $1 \le \nu_0 \le \frac{\log(N+4\sqrt{N}+1)}{\log l}$. Let

$$\bar{P} = (P, P'), \quad \bar{Q} = (Q, Q'), \quad \text{where} \quad \nu_l(\bar{P}) \ge \nu_l(\bar{Q}) \qquad (5.1)$$

Without loosing the generality we may assume that $\nu_l(P') \le \nu_l(P) := \nu$ and $\max(\nu_l(Q), \nu_l(Q')) \le \nu$. Multiplying the point $\bar{P}$ by E_r/l^ν we obtain for some $r \in \{p, q\}$ the point $\bar{R}$ of local orders $\text{ord}R = l^\nu$ and $\text{ord}R' \mid l^\nu$. To simplify the notation we denote the resulted point $\bar{R}$ by $\bar{P}$ in the following algorithm.

Algorithm $\mathcal{A}_0$

Input:

(1) Real number $\vartheta > 1$, semiprime $N \ge N_0(\vartheta)$ such that $p < q < \vartheta p$ and $B = B(N) < D = D(N)$ and E_r is $D-$smooth number while $B < D$ and $E_r = s_B(E_r)\bar{q}$, where $\bar{q} \le D^\sigma$.
(2) Elliptic curve $E = E(\mathbb{Q})$ such that the reduced curve $E(\mathbb{Z}_r)$ is cyclic group and has $D-$smooth order.
(3) Two finite points $\bar{P}, \bar{Q} \in E(\mathbb{Q})$ satisfying the condition

$$\text{ord}P' \mid \text{ord}P = l^\nu \qquad (5.2)$$

$$\text{ord}Q \mid l^\nu \quad \text{and} \quad \text{ord}Q' \mid l^\nu \qquad (5.3)$$

Output:
Decomposition $N = pq$ or the proof that $G_{\{\bar{P}, \bar{Q}\}} = G_{\{\bar{P}\}}$ is cyclic.

5.2   Brief Analysis of Algorithm $\mathcal{A}$

The algorithm is based on the consideration of the equation

$$\bar{k}\bar{P} = \bar{k}_0 + \bar{k}_1 l + ... + \bar{k}_{\nu-1}l^{\nu-1}\bar{P} = \bar{Q}, \tag{5.4}$$

on the elliptic curve E over $\mathbb{Z}_N$. We try to compute subsequently the coefficients $\bar{k}_0, \bar{k}_1, ..., \bar{k}_{\nu-1} \in [0, l)$ of the representation of $\bar{k}$ in the base l and if in some step the computation fails then we deduce that the l–adic local orders $\nu_l(\text{ord}P)$ and $\nu_l(\text{ord}P')$ must be distinct for a given prime l, and this could be discovered in deterministic time $t_{\mathcal{A}} = |\bar{X}|(B^2 + D^{3/2})^{1+o(1)}$ if the final algorithm $\mathcal{A}$ runs over all possible values ($l \in (B, D]$) and points $\bar{P} \in \bar{X}$. Otherwise (that is if sall computations of $\bar{k}_i$ were successful) we find the point $\bar{P} \in \bar{X}$ which generates the group $G_{\bar{X}}$.

The algorithm runs along the following steps

(1) Let $0 \le \bar{k} < l^\nu$, $\bar{k} = \sum_i \bar{k}_i l^i$ with $k_i \in [0, l-1]$ and considering the equality 5.4 above we pass to the next step
(2) Try to compute $\bar{k}_0$ satisfying the condition

$$\bar{k}_0(l^{\nu-1}\bar{P}) = l^{\nu-1}\bar{Q} := l^{\nu-1}\bar{R}_0$$

(3) For each $\bar{k}_0 \in [0, l)$ check if $\nu_l(\text{ord}\bar{P}_r) = \nu_l(\text{ord}\bar{P}_{r'})$ computing the multiple point $l^{\nu-1}\bar{P}$. If in the computation we obtain the nontrivial divisor $d \mid N$ ten halt and return d as output. Otherwise go to next step.
(4) Try to find $k_1 \in [0, l)$ satisfying the equality

$$\bar{k}_1(l^{\nu-1}\bar{P}) = l^{\nu-2}(\bar{Q} - k_0\bar{P}) := l^{\nu-2}R_1, \tag{5.5}$$

$k_2 \in [0, l)$ satisfying the equality

$$\bar{k}_2(l^{\nu-1}\bar{P}) = l^{\nu-3}(\bar{Q} - (k_0 + k_1 l)\bar{P}) := l^{\nu-3}R_2, \tag{5.6}$$

and so on.
(5) If computing $\bar{k}_i$ for $i = 1, 2, ..., \nu - 1$ we did not factor N then we halt and return $\bar{k} = \sum_i \bar{k}_i l^i$ satisfying $\bar{k}\bar{P} = \bar{Q} \in E(\mathbb{Z}_N)$ that is that the algorithm returns $G_{\bar{X}} = G_{\{\bar{P}\}}$ as ouput.

If the group $G_{\{\bar{P}, \bar{Q}\}}$ is cyclic then the order of it is equal to lcm of orders ord$\bar{P}$ and ord$\bar{Q}$ equal to l^ν by the assumptions and $\bar{P}$ is a generator of this group, hence $\bar{Q} = \bar{k}\bar{P}$ for some integer k. Therefore the output of the above algorithm given in step 5 above implies that if $|G_{\bar{X}}| > N^{1-\eta}$ then in deterministic time $|\bar{X}|$ we find a point $\bar{P} \in \bar{X}$ such that $\text{ord}(\bar{P}) > N^{1-\eta}$.

Otherwise there must exist in the set $\bar{X}$ some (E, D)–separating witness. This witness is recognized by the algorithm PHZ in the first time when it fails. We can use this witness to decompose $N = pq$ in deterministic time $t_{\mathcal{A}}$.

We consider the reductions of the above equation for $P_r = \pi_r(\bar{P})$ and $Q_r = \pi_r(\bar{Q})$ respectively for $r \in \{p, q\}$, where the point $\bar{Q}$ is finite. Then we have

$$(k_0 + k_1 l + ... + k_{\nu-1} l^{\nu-1}) P_r = Q_r, \tag{5.7}$$

and similarly

$$(k'_0 + k'_1 l + ... + k'_{\nu-1} l^{\nu-1}) P_{r'} = Q_{r'}. \tag{5.8}$$

Now rising both sides to power $l^{\nu-1}$ we have that

$$k_0 (l^{\nu-1} P_r) = l^{\nu-1} Q_r := l^{\nu-1} R_0$$

$$k_1 (l^{\nu-1} P_r) = l^{\nu-2} (Q_r - k_0 P_r) := l^{\nu-2} R_1,$$

and so on, where k_i are the minimal nonegative integers which are congruent to $\bar{k}_i$ modulo the order of the point $l^{\nu-1} P_r$. Since $k_i \in [0, l-1]$ we have by the assumptions (5.2) and (5.3) that the order of $\text{ord}(l^{\nu-1} P_r)$ is l, hence $k_i \leq \bar{k}_i$ must be equal to o $\bar{k}_i$ for $i = 0, 1, ..., \nu-1$. Similarly the order of point $l^{\nu-1} P_{r'}$ divides l so can be either l or 1. In the former case we argue as above to conclude that $k'_i = k_i = \bar{k}_i$ for $i = 0, 1, ..., \nu-1$. The latter case means that the point $l^{\nu-1} P_{r'}$ is not finite, which means that $\nu_l(l^{\nu-1} P_{r'}) = 0$, hence the local orders $\text{ord} P_r$ and $\text{ord} P_{r'}$ are divisible by different powers of l and this can be recognized1 by computing $l^{\nu-1} \bar{P}$ that is in $\log(l^{\nu})$ operations addition operations in $E(\mathbb{Z}_N)$ which gives the determinstic complexity $O(\log N \log B \log^2 N) = O(\log^4 N)$. Since the number of k_i is $\nu \leq \log N$ and $k_i < l \leq D$ the application of Shanks, method allow to reduce the complexity to $D^{1/2+o(1)}$. This completes the proof of the lemma.

Remark: The deterministic comlexity of the algorithm $\mathcal{A}_0$ is bouded by $O(l\nu \log(l^{\nu})) = O((\log N)^2 (D \log D))$ additions in $E(\mathbb{Z}_N)$. The algorithm $\mathcal{A}$ uses the algorithm $\mathcal{A}_0$ as subroutine and the fact that each D−smooth number is of type $m_D = m_B(\bar{q})$, and the squarefree kernel $(\bar{q})^* \leq D^{\sigma}$. In the algorithm $\mathcal{A}$ we run the loop over all possible multiple points $m_B \frac{E_r}{l^{\nu}} \bar{P} \in E(\mathbb{Z}_N)$ which are simulating by the multiple points $\left(m_B \frac{\bar{q}^{\kappa}}{l^{\nu}} \right) \bar{P} = \left(\frac{\bar{q}^{\kappa}}{l^{\nu}} \right) \bar{R}$, for a given finite poin $\bar{R}$.

If the above assumptions hold and neither $\bar{P}$ nor $\bar{Q}$ is (E, D)−witness then if $N, E, P, Q, l \in (B, D]$ and $\nu \geq 1$ satisfy the above conditions and if $G_{\bar{X}} \in E(\mathbb{Z}_N)$ is noncyclic. Then applying Algorithm $\mathcal{A}_0$ one can find the decomposition $N = pq$ in deterministic time $O(D^{1/2+o(1)})$ (provided $D \geq \log^C N$ for sufficiently large $C > 0$) as output.

Proof. By the asumption we have that $E(\mathbb{Z}_N) \simeq E(\mathbb{F}_p) \times E(\mathbb{F}_q)$ is a direct sum of two cyclic groups. The idea is to use the Silver-Pohlig-Hellman algorithm adopted to factoring of N applied in [10] for the case of modular group $\mathbb{Z}_N^*$. By the conditions 5.2 and 5.3 we have that if $< \bar{P}, \bar{Q} >_{E(\mathbb{Z}_N)}$ is cyclic then Q must belong to the subgroup in $E(\mathbb{Z}_N)$ generated by Q, that is there should exists $1 \leq k'' \mid l^{\nu}$ such that

$$\bar{k} \bar{P} = \bar{Q}, \quad \text{where} \quad \bar{P}, \bar{Q} \in E(\mathbb{Z}_N).$$

On the other hand if it is not the case then the above algorithm $\mathcal{A}_0$ returs the nontrivial divisor $d \mid N$ as output.

Namely we consider the reductions of the above equation for $P_r = \pi_r(\bar{P})$ and $Q_r = \pi_r(\bar{Q})$ respectively for $r \in \{p, q\}$, where the point $\bar{Q}$ is finite. Then we have

$$(k_0 + k_1 l + \dots + k_{\nu-1}l^{\nu-1})P_r = Q_r, \tag{5.9}$$

and similarly

$$(k'_0 + k'_1 l + \dots + k'_{\nu-1}l^{\nu-1})P_{r'} = Q_{r'}. \tag{5.10}$$

Now rising both sides to power $l^{\nu-1}$ we have that

$$k_0(l^{\nu-1}P_r) = l^{\nu-1}Q_r := l^{\nu-1}R_0$$

$$k_1(l^{\nu-1}P_r) = l^{\nu-2}(Q_r - k_0 P_r) := l^{\nu-2}R_1,$$

and so on, where k_i are the minimal nonegative integers which are congruent to $\bar{k}_i$ modulo the order of the point $l^{\nu-1}P_r$. Since $k_i \in [0, l-1]$ we have by the assumptions 5.2 and 5.3 that the order of $\mathrm{ord}(l^{\nu-1}P_r)$ is l, hence $k_i \leq \bar{k}_i$ must be equal to o $\bar{k}_i$ for $i = 0, 1, ..., \nu-1$. Similarly the order of point $l^{\nu-1}P_{r'}$ divides l so can be either l or 1. In the former case we argue as above to conclude that $k'_i = k_i = \bar{k}_i$ for $i = 0, 1, ..., \nu-1$. The latter case means that the point $l^{\nu-1}P_{r'}$ is not finite, which means that $\nu_l(l^{\nu-1}P_{r'}) = 0$, hence the local orders $\mathrm{ord}P_r$ and $\mathrm{ord}P_{r'}$ are divisible by different powers of l and this can be recognized1 by computing $l^{\nu-1}\bar{P}$ that is in $\log(l^\nu)$ operations addition operations in $E(\mathbb{Z}_N)$ which gives the deterministic complexity $O(\log N \log B \log^2 N) = O(\log^4 N)$. Since the number of k_i is $\nu \leq \log N$ and $k_i < l \leq D$ the application of Shanks, method allow to reduce the complexity to $D^{1/2+o(1)}$. This completes the proof of the lemma.

6 Counting Function $\psi_{\beta,\sigma}(I, \beta, D)$ and Frequences of Admissible Curves

Definition 6. *Let $I = [I^-, I^+]$ be an interval. Let $\psi_\beta(I, B, D)$ be the number of D–smooth numbers $m \in I$ such that the maximal B–smooth factor of m is $\geq (I^+)^\beta$. Every D–smooth number m such that $s_B(m) \geq m^\beta$ is called (B, β, D)–smooth number.*

Clearly the number of (B, β, D)–smooth numbers is at least as large as $\psi_\beta(I, B, D)$. Analogously let $\psi_{\beta,\sigma}(I, B, D)$ count the number of $m's$ counted by the function $\psi_\beta(I, B, D)$ that satisfy the additional condition

$$(m/s_B(m)^* \leq D^\sigma, \quad (\sigma \in [1, 2]).$$

Every (B, β, D)–smooth number satisfying the last condition is called (B, β, D, σ)– admissible number. Clearly the number of (B, β, D, σ)– admissible numbers is at least as large as $\psi_\beta(I, B, D)$.

In what follows we and $B = B(N), D = D(N)$.

6.1   Elliptic Curve Analog and Decomposition $N = pq$

We focus on the significant case from the practical point of view when $N = pq$, $(p < q)$ is the semiprime number []. Let $E = E_{\bar{b}}(\mathbb{Z}_N)$ and $Q = (x, y) \in E_{\bar{b}}(\mathbb{Z}_N)$ be chosen by the random selection of triples $T = (x, y, b_1) \in \mathbb{Z}_N^3$ (according to the given pseudorandom generator) such that $\Delta_{\bar{b}} = 6(4b_1^3 + 27b_2^2)$ is coprime to $N = pq$ $(p < q)$. Let $b_2 := y^2 - x^3 - b_1 x$ and $\bar{b} := (b_1, b_2) \in \mathbb{Z}_N^2$. For elliptic curve E over $\mathbb{Z}_N$ and $r \in \{p, q\}$ we let $E_r := E(\mathbb{F}_r)$ be the order of ellitic curve E over finite field $\mathbb{F}_r$ and $\mathrm{ord}Q_r$ be the order of the related reduction of Q modulo r. Let I_r be the interval $I_r = (r + 1 - \sqrt{r}, r + 1 + \sqrt{r})$. For the given smoothness parameter $B = B(N)$ we let

$$f_r(B) = \frac{\#\{s \in I_r \text{ such that } s \text{ is } B - \text{smooth}\}}{\lfloor 2\sqrt{r} \rfloor + 1}$$

expressing the probability that a random integer from the interval I_r is $B-$smooth number.

Let $B = L(\alpha, r)$ be the familiar subexponential function. In the problem of decomposition of positive integer N (see [6]) the author proves that the number of trials for T, needed to obtain elliptic curve with $B-$smooth local order for some $r \in \{p, q\}$ is $\Theta(1)/f_p(B)$, while in order to decompose $N = pq$ a bit worse evaluation, equal to $\frac{\log p}{f_p(B)} = L\left(\frac{1 + o(1)}{2\alpha}, p\right)$ as N tends to infinity is proved under the plaussible conjecture concerning the distibution od $B-$smooth numbers in the interval I_r.

On the other hand the cost of coputing $m_B(Q)$ is $L(\alpha + o(1), p)$, where m_B is the product of all primes $l \leq B$ in powers not exceeding $\frac{\log(I_r^+)}{\log 2}$. This gives the total complexity of the related algorithm equal to $L(\frac{1}{2\alpha} + \alpha + o(1), p) = L\left(\sqrt{2} + o(1), p\right)$ as N tends to infinity, by the optimal choice $\alpha = \frac{1}{\sqrt{2}}$ (see [6].

6.2   Subexponential Values of B and D

Let $B = L(\alpha, r) < D = L(\alpha_0, r)$, where $\alpha_0 = 1/\sqrt{2}$. The analysis of the distribution of $D-$smooth local orders $E_r := \#E(\mathbb{F}_r)$ such that $s_B(E_r) \geq E_r^\beta$ gives the further insight into the better understanding and further investigation of the complexity of factorization algorithm for $N = pq$ in terms of the ratio D/B under the analogous conjecture concerning the $(B, \beta, D, \sigma)-$admissible numbers E_r.

In the related algorithm the $B-$smooth multiples $m_B Q \in E(\mathbb{Z}_N)$ are computed until the related point $Q_r \in E(\mathbb{F}_r)$ is not finite and then we apply the method described briefly below.

Namely performing $B^{2+o(1)}$ adding operation in $E(\mathbb{Z}_N)$ one can find the decomposition $N = pq$ provided E_r is $B-$smooth number. However if E_r is $(B, \beta, D, \sigma)-$admissible number then the suitable search of suitably chosen multiples $\prod_q q^{\nu_q} m_B(Q) \in E(\mathbb{Z}_N)$ gies the decomposition $N = pq$ in deterministic time $D^{\sigma + o(1)}$ resulting in the decomposition of N in deterministic time

$t = (B^2 + D^\sigma)^{1+o(1)}$. Assumint the related β–Conjecture we derive then the average-case time complexity $t \leq L(2\alpha_0 + o(1), p)$, which is stronger than the one obtained in [6] provided the condition 6.1 below holds true.

6.3 Average-Case Time Complexity of Decomposition $N = pq$

The main results on E-separating and (E, γ)-nonseparating witnesses are based on Proposition 2, the β-Conjecture and the following lemmas.

Given the proper pair (E, Q) we are able to decompose $N = pq$ in deterministic time $(B^2 + D^\sigma)^{1+o(1)}$, which improves the estimate following by Proposition 2 (see [9]). If the analogous conjecure for (B, β, D)–smooth numbers as for B–smooth numbers assumed in [6] is true then we get the following result on the deterministic time of decomposition $N = pq$ for subexponential values of B and D.

Theorem 6.1. *Assume that $N = pq$, where $p < q < \vartheta p$, and $B = L(\alpha_0/\lambda, r)$, $D = L(\alpha_0, r)$, $\alpha_0 = 1/\sqrt{2}$. Let $\sigma, \lambda \in [1, 2]$, $\beta \in [0, 1)$ and $\theta = \theta_\sigma$ be the related constant occuring in $\beta-$ Conjecture. If (E, Q) is a proper pair then we have the following average-case time complexity of decomposition $N = pq$*

$$t = L\left(\frac{1 - \theta_\sigma(1 - \beta)}{\sigma\alpha_0} + \sigma\alpha_0 + o(1),\ r\right).$$

Hence comparing it with the average-case time complexity $t = L(2\alpha_0 + o(1), r)$ given in [6] we obtain the nontrvial result provided

$$\theta_\sigma \geq \frac{2 - 2\sigma + \sigma^2}{2(1 - \beta)}. \tag{6.1}$$

The best lower bound equal to $1/(2(1 - \beta))$ is attained for $\sigma = 1$.

Proof. The average time for selection of triple $T = (x, y, b_1) \in \mathbb{Z}_N^3$ such that E_r is $(I_r, B, \beta, D, \sigma; 0)$-admissible number is in view of β-Conjecture equal to

$$L\left(\frac{1 - \theta_\sigma(1 - \beta)}{2(\alpha_0/\lambda)} + o(1),\ r\right) = L\left(\frac{1 - \theta_\sigma(1 - \beta)}{2\alpha_0}\lambda + o(1),\ r\right)$$

Moreover in view of Proposition 1 and Proposition 2 above one can compute the related multiple $s_B(E_r)\frac{s_p}{s_B}$ in deterministic time $B^2 + D^{\sigma+o(1)}$ which implies that the total complexity of factorization $N = pq$ is equal to

$$t = L\left(\frac{1 - \theta_\sigma(1 - \beta)}{2\alpha_0}\lambda + o(1),\ r\right) \max\left(L\left(\frac{2\alpha_0}{\lambda} + o(1),\ r\right),\ L(\sigma\alpha_0 + o(1), r)\right) \tag{6.2}$$

$$= L\left(\frac{1 - \theta_\sigma(1 - \beta)}{2\alpha_0}\lambda + \max\left(\frac{2\alpha_0}{\lambda}, \sigma\alpha_0\right) + o(1),\ r\right),$$

Now choosing λ satisfying $2\alpha_0/\lambda = \sigma\alpha_0$, hence $\lambda = 2/\sigma$ and we obtain the inequality

$$t = L\left(\frac{1 - \theta_\sigma(1-\beta)}{\sigma\alpha_0} + \sigma\alpha_0 + o(1),\ r\right) = L(\phi(\beta,\sigma) + o(1),\ r)$$

where $\phi(\beta,\sigma) = \frac{1-\theta_\sigma(1-\beta)}{\sigma\alpha_0} + \sigma\alpha_0 \leq 2\alpha_0$, provided $\theta_\sigma \geq \frac{2-2\sigma+\sigma^2}{2(1-\beta)}$, as required.

6.4   Algorithm $\mathcal{A}$

The algorithm $\mathcal{A}$ reduces the search for the (E, D)−decompsition witness for N to Algorithm $\mathcal{A}_0$. Namely we start from the first two points $\bar{P}_1 = \bar{P}, \bar{P}_2 = \bar{Q}$ and assume that $\nu_l(\mathrm{ord}Q) \mid \nu_l(\mathrm{ord}P) := \nu$, where $\mathrm{ord}P \mid E(\mathbb{Z}_r)$ and $E(\mathbb{Z}_r)$ is cyclic, E_r is D−smooth and known.

We consider the multiple points

$$\frac{E_r}{l^{\nu_l(\mathrm{ord}E_r)}}P \quad \text{and} \quad \frac{E_r}{l^{\nu_l(\mathrm{ord}E_r)}}Q \tag{6.3}$$

respectively for some $l \in (B, D]$, $\nu \leq \nu_l(E_r)$ such that $\nu_l(\mathrm{ord}Q) \leq \nu_l(\mathrm{ord}P) \leq \nu_l(E_r)$, hence they satisfy the conditions required in algorithm $\mathcal{A}_0$. If all local orders $\mathrm{ord}P$ and $\mathrm{ord}Q$ have the same D−smooth divisors then the related ν_l are equal and therefore we cal find the solution $k\bar{P} = \bar{Q}$ using the above algorithm. For the estimation of the complexity we need the following:

Let $m* = \prod_i q_i$, where $q_1 > q_2 > \ldots > q_k$. We say that the sequence $(q_1, ..., q_k)$ belongs to the tuple $\bar{l} = (l_1, ..., l_k)$ if $q_i \in [2^{l_i}, 2^{l_i+1})$ for $i = 1, 2..., k$.

Definition 7. *The tuple $\bar{l} = (l_1, l_2, ..., l_k)$ is called Δ−admissible if*

$$l_1 \geq l_2 \geq \ldots \geq l_k$$

and

$$\sum_{i \leq k} l_i \leq \log \Delta / \log 2 \tag{6.4}$$

We have (see Lemma 2 [9]) allowing to reduce counting the sequences $(q_1, ..., q_k)$ with coordinates depending only on the values of l_i, $i = 1, 2..., k$ of Δ−admissible tuples $\bar{l}$.

Lemma 2. *The number of sequences $(q_1, ..., q_k)$ belonging to a fixed Δ−admissible tuple $\bar{l}$ is $\leq \Delta$. Moreover the number of Δ−admissible tuples $(l_1, ..., l_k)$ to which some sequence may belong is $O(\log \Delta)^k$, where the constant implied by the symbol O does not depend on Δ.*

Remark: The deterministic comlexity of the algorithm $\mathcal{A}_0$ is bouded by $O(l\nu \log(l^\nu)) = O((\log N)^2(D \log D))$ additions in $E(\mathbb{Z}_N)$. The algorithm $\mathcal{A}$ uses the algorithm $\mathcal{A}_0$ as subroutine and the fact that each D−smooth number is of type $m_D = m_B(\bar{q})$, and the squarefree kernel $(\bar{q})^* \leq D^\sigma$. In the algorithm $\mathcal{A}$ we run the loop over all possible multiple points $m_B \frac{E_r}{l^\nu}\bar{P} \in E(\mathbb{Z}_N)$ which are simulating by the multiple points $\left(m_B \frac{\bar{q}^\kappa}{l^\nu}\right)\bar{P} = \left(\frac{\bar{q}^\kappa}{l^\nu}\right)\bar{R}$, for a given finite point $\bar{R}$.

References

1. Coppersmith, D.: Finding a small root of a bivariate integer equation; factoring with high bits known. In: Maurer, U. (ed.) EUROCRYPT 1996. LNCS, vol. 1070, pp. 178–189. Springer, Heidelberg (1996). https://doi.org/10.1007/3-540-68339-9_16
2. Dieulefait, L.V., Urroz, J.: Factorization and malleability of RSA moduli, and counting points on elliptic curves modulo N. Mathematics **8**(12), 2126 (2020)
3. Dryło, R., Pomykała, J.: Integer factoring problem and elliptic curves over the ring $\mathbb{Z}_n$. In: Colloquium Mathematicum, vol. 159, pp. 259–284. Instytut Matematyczny Polskiej Akademii Nauk (2020)
4. Dryło, R., Pomykała, J.: Smooth factors of integers and elliptic curve based factoring with an oracle. Banach Center Publ. **126**, 73–88 (2023)
5. Lenstra, H.W., et al.: Elliptic curves and number-theoretic algorithms. Universiteit van Amsterdam Mathematisch Instituut (1986)
6. Lenstra Jr, H.W.: Factoring integers with elliptic curves. Ann. Math. 649–673 (1987)
7. Pohlig, S., Hellman, M.: An improved algorithm for computing logarithms over GF (p) and its cryptographic significance (corresp.). IEEE Trans. Inf. Theory **24**(1), 106–110 (1978)
8. Pomykała, J., Jurkiewicz, M., Żołnierczyk, O., Prabucka, K.: Enhanced Performance of ECM for RSA Modulus1 via Generealized B-smoothness. In: Send to ISAAC 2024 (2024)
9. Pomykaa, J., Onierczyk, O.: Elliptic curve-integer factorizatiin and witnesses. In: Accepted for the conference ICCS 24 (2024)
10. Żrałek, B.: A deterministic version of Pollard's $p-1$ algorithm. Math. Comput. **79**(269), 513–533 (2010)

Statistically Secure Multiparty Computation of a Biased Coin

Amir Zarei[(✉)] and Staal A. Vinterbo

Norwegian University of Science and Technology, Gjøvik, Norway
{amir.zarei,staal.vinterbo}@ntnu.no

Abstract. Secure multiparty computation (MPC) enables multiple parties to jointly evaluate functions while ensuring the privacy of their inputs. Sampling a biased coin is an important MPC building block for evaluating randomized functions. This paper presents a new MPC protocol for sampling a biased coin using $2d+1$ unbiased coins. The protocol is statistically secure against passive adversaries and can be implemented using $11d + 5$ multiplications and five rounds. Here, d is associated with the used finite field size p as $\lceil \log_2 p \rceil = 2d + 1$. The protocol is based on secure arithmetic in $\mathbb{Z}_p$ and can be implemented using any linear secret-sharing scheme. Active security for this protocol can be achieved by incorporating additional existing protocols. The proposed protocol offers significant reductions in round and communication complexities compared to a solution offered by Eriguchi et al. that requires $(5n + 19)d$ multiplications and 11 rounds for n parties.

Keywords: Secure Multiparty Computation · Secret Sharing · Biased Coin

1 Introduction

In recent decades, more and more data about us have been collected. With the recent breakthroughs in artificial intelligence, particularly in models trained on vast amounts of data (see, e.g., OpenAI's GPT [1] and Google's BERT [2]), the demand for data is ever-increasing. Unfortunately, as more data about us is collected, ensuring its security becomes even more important. An underutilized tool in this context is secure multiparty computation (MPC), which enables a group of mutually distrusting parties to compute a function on their joint data while ensuring that each party's data remains undisclosed to the other parties. From a privacy perspective, this is particularly powerful when combined with differential privacy [3] as it not only conceals each party's data from the others but also from any external entity [4]. However, one of the main barriers to adopting MPC and such a combination is the relative inefficiency of MPC compared to simply sharing the data. Our research aims to address and mitigate this challenge.

© The Author(s), under exclusive license to Springer Nature Switzerland AG 2025
F. Martinelli and R. Rios (Eds.): STM 2024, LNCS 15235, pp. 66–81, 2025.
https://doi.org/10.1007/978-3-031-76371-7_5

An important class of MPC solutions provides secure arithmetic over a field or ring, e.g., $\mathbb{Z}_p$ for some prime p, relying on secret sharing. The general framework of these solutions is similar—by abstracting away the details of the constructions, secure arithmetic forms the basis for the design of MPC protocols. This idea was formally established with the concept known as the Arithmetic Black Box (ABB) proposed by Damgård et al. [7].

The generation of a biased coin is a fundamental MPC building block. It plays a pivotal role in implementing randomized algorithms such as Monte Carlo simulations and differentially private algorithms [9]. In particular, Dwork et al. [4] demonstrate that sampling from binomial, Poisson, and geometric distributions can be simplified by sampling biased coins. Our contribution lies in reducing the complexities associated with communication and rounds required for computing a shared biased coin. This advancement offers a more efficient primitive for generating random samples from discrete probability distributions. Such a practice has become increasingly important in the domain of differential privacy, as addressed by Canonne et al. [5] and the differential privacy team at Google [6].

1.1   Further Background and Previous Research

In our discussion, a coin with bias α is formally defined as a random bit with a probability α of taking the value 1. Dwork et al. [4] introduce a simple and pioneering method for secure multiparty sampling of a coin with bias α. This method involves discrete sampling of d unbiased coins and using them as the binary representation of a uniformly random number r. By comparing the values of r and α, the desired coin with bias α can be obtained. Dwork et al. [4] also describe how this method can be implemented using multiplications of $O(d)$.

The study of Eriguchi et al. [11] presents a secure implementation of the differentially private geometric mechanism for $n > 2$ parties. The foundation of their method relies on employing secret sharing over the finite field $\mathbb{Z}_p$. As a core building block of this protocol, they present a technique for sampling a shared biased coin that depends on generating d unbiased coins, as described by Dwork et al. Their proposed biased coin protocol requires $(5n+19)d$ multiplications in 11 rounds [11,28]. As the geometric distribution quantifies the probability of achieving the first success by taking a certain number of biased coins, denoted N, their protocol to generate a geometric sample requires $(5n + 19)dN + 17N + 5n + 3$ multiplications. This example highlights the significance of reducing the complexity involved in generating a single biased coin, as it acts as a multiplicative factor in the total communication complexity.

Zarei et al. [10] provide an MPC protocol for privately generating a shared biased coin with fewer multiplication protocol invocations and rounds than the solution by Eriguchi et al. [11,28]. However, this protocol assumes a specific assumption for the field size p, namely that p is a Mersenne prime or similar primes approximating powers of 2. This requirement limits the protocol's general applicability in various scenarios.

1.2   Our Result

Recall that a coin with bias α can be sampled by comparing a uniform random variable r to α. Our research introduces an MPC protocol for sampling such a coin. Inspired by Zarei et al. [10], our technique reduces the comparison of r with α to the least significant bit extraction task, which is cheaper regarding round and communication complexities. The proposed protocol is statistically secure and applies to any linear secret sharing scheme over $\mathbb{Z}_p$. This protocol can be realized assuming that the field size p is sufficiently large, i.e., $p > 2^{2d}$ for $d = \kappa + \log n$, where κ is the security parameter and n is the number of parties. We show that if the number of bits required to represent p is $\lceil \log_2 p \rceil = 2d + 1$, i.e., $2^{2d} < p \leq 2^{2d+1}$, the protocol can be implemented using only $11d + 5$ multiplications in 5 rounds. For any $d \geq 1$, such a p exists by Bertrand's postulate.

In contrast to the protocols proposed by Eriguchi et al. [11] and Zarei et al. [10], perfect security cannot be provided, even if the underlying primitives guarantee this. The security of our protocol is statistical, which still allows it to be unconditional. For this protocol, active security is not directly obtained from the active security of its primitives. Still, it can be achieved when the parties prove that a provided input is less than some public bound through existing additional protocols.

While guaranteeing slightly weaker security, our proposed protocol reduces both round and communication complexities, compared to the solution offered by Eriguchi et al. [11] that needs $(5n + 19)d$ multiplications in 11 rounds (we will discuss this in more detail in Sect. 5). The enhanced efficiency of our protocol suggests potential advancements in the combination of MPC and differential privacy, particularly in scenarios that require noise generation from discrete probability distributions. Compared to the solution proposed by Zarei et al. [10], our protocol removes the restrictive assumption that the field size must be a Mersenne prime or similar by adopting a different method to extract the least significant bit. This adaptation broadens the applicability of our protocol across a wider range of field sizes, making it more versatile in various cryptographic scenarios.

2   Secure Arithmetic Notation and Primitives

We present our protocol within the framework of secure MPC, specifically focusing on the linear secret sharing scheme used in the ABB model [7]. The ABB model provides a foundation for secure computation, enabling multiple parties to perform secure arithmetic operations on shared secrets or encrypted data. It is described as an ideal functionality within the universal composability framework by Canetti [8]. Our proposed protocol is compatible with any schemes that realize this functionality. For example, combining Shamir's secret sharing scheme [13] with the protocols presented by Ben-Or et al. [14] can achieve this functionality under passive, active and adaptive security conditions. We employ the ABB model to abstract the complex underlying cryptographic operations, thereby obviating the need for individual security proofs for each operation within our protocol. The use of the ABB model, a common practice in MPC literature, allows security to be effectively derived from the model's established security, thus streamlining the development of secure protocols by leveraging a proven, abstract model.

2.1 The Arithmetic Black-Box (ABB)

In this work, ABB enables n parties, $P_1, P_2, \ldots, P_n$, to securely store and retrieve secret values from the finite field $\mathbb{Z}_p$ and to execute arithmetic operations. We assume that secure storage (input/output) is implemented as linear secret sharing and that p is a prime. $[\![a]\!]$ denotes the shared value of a among the parties $P_1, P_2, \ldots, P_n$. The notation $[\![a]\!]_B$ will be used to refer to a bit-decomposed value of a, i.e. it is shorthand for the sequence $([\![a_{l-1}]\!], [\![a_{l-2}]\!], \ldots, [\![a_0]\!])$, where each $[\![a_i]\!]$ represents the i-th bit of $[\![a]\!]_B$ and $[\![a]\!] = \sum_{i=0}^{l-1} [\![a_i]\!] 2^i$. We define l as the number of bits required to represent any value in $\mathbb{Z}_p$, calculated as $l = \lceil \log_2 p \rceil$. ABB provides the following operations:

- **Input:** A party P_i may input a value $a \in \mathbb{Z}_p$ by sharing it among the parties.
- **Output:** The parties reconstruct and publicly reveal a shared secret $[\![a]\!]$, making the value a known to all parties.
- **Linear combination:** The parties can calculate a linear combination of secret shares using the operation $[\![\sum_i \alpha_i a_i]\!] \leftarrow \sum_i \alpha_i [\![a_i]\!]$, exploiting the inherent linear properties of the secret sharing scheme in use.
- **Multiplication:** The parties may compute the product of two secret shares, expressed as $[\![ab]\!] \leftarrow [\![a]\!][\![b]\!]$. This multiplication protocol requires interaction among the parties to ensure no leaks of secure data.
- **Addition:** The parties execute the addition of two stored secret shares, expressed as $[\![a+b]\!] \leftarrow [\![a]\!] + [\![b]\!]$.
- **Addition with a public constant:** The parties perform the addition of a shared value $[\![a]\!]$ with a public constant c, expressed as $[\![a+c]\!] \leftarrow [\![a]\!] + c$.
- **Multiplication by a public constant:** The parties compute the multiplication of a shared value $[\![a]\!]$ by a public constant c, expressed as $[\![ca]\!] \leftarrow c[\![a]\!]$.

Moreover, XORing two shared bits $[\![a]\!]$ and $[\![b]\!]$ can be computed by using one invocation of the multiplication protocol, expressed as $[\![a]\!] \oplus [\![b]\!] = [\![a]\!] + [\![b]\!] - 2[\![a]\!][\![b]\!]$. The notation $[\![a]\!]_b$ is a concise representation for l uniformly random shared bits $([\![a_{l-1}]\!], [\![a_{l-2}]\!], \ldots, [\![a_0]\!])$, where $[\![a_i]\!]$ represents the i-th bit of $[\![a]\!]_b$. Furthermore, a publicly known value α can be decomposed into bits denoted as α_B. Each bit α_i corresponds to the i-th bit of α_B, with α_0 representing the least significant bit. We note that, considering $[\![a]\!]_b$ and α_B, XORing bits in their j-th bit position is represented as $\alpha_j \oplus [\![a_j]\!] = \alpha_j + [\![a_j]\!] - 2\alpha_j[\![a_j]\!]$ (where the subscripts B and b have been omitted for clarity).

The complexity analysis will address rounds and communication complexities. This implies that secure computation of linear combinations follows directly from the linearity of the underlying scheme and is therefore considered without cost. The number of sequential invocations of the multiplication protocol represents the round complexity. It is assumed that the ABB allows multiplication invocations to be carried out concurrently and are treated as a single round. The total number of invocations of the multiplication protocol determines the communication complexity, which is commonly used as a measure of the total volume of data transmitted by the parties. In addition to this, in line with related works [15–17], we equate the complexity of sharing from every party to one multiplication invocation in one round. This is because the multiplication protocol usually requires at least one input from each party.

Regarding security, let $\text{VIEW}_\pi(P_i) = (\text{in}_i, r_i, M)$ denote the view of party P_i including its input in_i, its internal random coin tosses r_i, and the messages M exchanged between parties during the execution of protocol π. Let $I = \{P_{i_1}, ..., P_{i_t}\}$ denote a subset of the parties for $t < \frac{n}{2}$, $\text{VIEW}_\pi(I)$ denote the combined view of parties in I during the execution of protocol π. We say that a protocol π is statistically secure in the presence of passive adversaries if, for each coalition of parties of size at most $t < \frac{n}{2}$, there exists a simulator S_I such that the distributions of $\{\text{VIEW}_\pi(I), \text{out}_I\}$ and $S_I(\text{in}_I, \text{out}_I)$, where $\text{in}_I = \cup_{P_i \in I}\{\text{in}_i\}$ and $\text{out}_I = \cup_{P_i \in I}\{\text{out}_i\}$, have a total variation distance exponentially small in some security parameter κ. Following Damgård et al. [20], we consider π to be perfectly secure if the two above distributions are identical except with negligible probability of order p^{-1}. This failure event does not compromise privacy and refers to the inability to generate an appropriate random value that is detected. In other words, the protocol might terminate with some abort symbol with a negligible probability in the security parameter κ as $p > 2^\kappa$. We emphasize that, similar to previous works, the security of our protocol is only shown in the ABB model of computation, meaning we are only concerned with primitives that securely realize this ideal functionality. This reduces security to ensure that inputs are proper, e.g., taken from some subset of $\mathbb{Z}_p$, and that no information leaks when values are output (revealed).

2.2   Complex Primitives

The protocol proposed in this work is not directly formulated within the ABB model. Instead, we provide a high-level outline of several primitives derived from previous research, which serve as sub-protocols in constructing our proposed protocol. Although these primitives might occasionally fail to generate random values, the probability of such failures is on the order of p^{-1}. Given the sufficiently large field size in use, this failure probability is considered negligible and can be disregarded. For a detailed discussion of this decision, see Damgrd et al. [20].

Random Element, Element Inversion, and Prefix Product. The parties have the ability to generate a uniformly random shared value, $[\![r]\!]$, by employing the RAN_p protocol introduced by Damgård et al. [20]. The RAN_p protocol involves each party P_i sharing a randomly chosen value r_i, and the sum of these values, $[\![\sum_{i=1}^n r_i \mod p]\!]$, yields a uniformly random shared value that remains unknown to all parties. The complexity of this protocol is equivalent to a multiplication protocol invocation in a round.

Damgård et al. [20] also provide a protocol to generate a random non-zero shared element $[\![r]\!]$ and its corresponding shared inverse $[\![r^{-1}]\!]$. This technique generates two random shared elements $[\![r]\!]$ and $[\![s]\!]$. The product rs is then computed and revealed. If it is equal to zero, the protocol fails. However, if rs is non-zero, $[\![r]\!]$ represents a non-zero shared element that remains unknown to the parties due to masking by $[\![s]\!]$. Moreover, the shared inverse $[\![r^{-1}]\!]$ can be computed without incurring any additional cost, as $[\![r^{-1}]\!] = (rs)^{-1}[\![s]\!]$. The complexity of this technique is determined by the complexity of two multiplications needed to generate $[\![r]\!]$ and $[\![s]\!]$, along with one more multiplication for their product. In total, three multiplications and two rounds are required to generate a random non-zero shared element and its shared inverse.

The inversion technique discussed above can be used to compute prefix products of k non-zero secret shared values using the method of Damgård et al. [20] with subsequent improvements by Reistad et al. [15–17]. Given $([\![a_1]\!], \ldots, [\![a_k]\!])$, the prefix products are defined as $[\![b_j]\!] = \prod_{i=1}^{j}[\![a_i]\!]$ for $j \in \{1, \ldots, k\}$. The implementation idea is first to set $r_0 = 1$ and generate k random non-zero shared elements $[\![r_i]\!]$, their inverses $[\![r_i^{-1}]\!]$, and the masks $[\![s_i]\!]$. Then, it computes $[\![r_{i-1}]\!][\![s_i]\!]$ for $i \in \{1, \ldots, k\}$ in parallel with the previous computation. Later, it masks each $[\![a_i]\!]$ and reveals the result as $m_i = [\![a_i]\!][\![r_{i-1}r_i^{-1}]\!]$ for $i \in \{1, \ldots, k\}$. We note that the computation of each $[\![r_{i-1}r_i^{-1}]\!]$ can be done without cost as $[\![r_{i-1}r_i^{-1}]\!] = [\![r_{i-1}s_i]\!] \cdot (r_i s_i)^{-1}$. Finally, the parties can compute the prefixes without cost as $[\![b_j]\!] = [\![r_j]\!] \cdot \prod_{i=1}^{j} m_i$ for $j \in \{1, \ldots, k\}$. Privacy of each $[\![a_i]\!]$ is guaranteed through the masking effect of the corresponding $[\![r_i^{-1}]\!]$. The efficient implementation of this method requires a total of $5k$ multiplications distributed over three rounds. More specifically, the preparation for the prefix product, including the generation of $[\![r_i]\!]$'s, $[\![r_i^{-1}]\!]$'s, $[\![s_i]\!]$'s and the multiplication of $([\![r_{i-1}]\!][\![s_i]\!])$'s can be accomplished using $4k$ multiplications during the first two rounds. Subsequently, m_i's can be obtained in the third round with additional k multiplications.

Random Bit(s). The parties have the ability to create a uniformly random shared bit $[\![b]\!] \in \{0, 1\}$ by the following RAN_2 protocol introduced in the work of Damgård et al. [20]. The parties first generate a random value $[\![r]\!]$ and reveal its square, r^2. The protocol fails if $r^2 = 0$; otherwise the parties compute a shared bit $[\![b]\!]$ as $[\![b]\!] = 2^{-1}((\sqrt{r^2}^{-1})[\![r]\!] + 1)$, where $0 \le \sqrt{r^2} \le \frac{p-1}{2}$. As for privacy, revealing r^2 does not disclose which of its two roots it contains. Moreover, the two possibilities occur with a probability of 2^{-1} each, thus the final bit $[\![b]\!]$ is uniformly random and unknown to all parties. The complexity of this method is bounded by the complexity of one multiplication for generating $[\![r]\!]$ and another multiplication for computing $[\![r^2]\!]$, which can be executed in two consecutive rounds.

As highlighted by Damgård et al. [20], the parties can extend the above approach to generate l uniformly random shared bits $[\![r]\!]_b = ([\![r_{l-1}]\!], \ldots, [\![r_0]\!])$

by l invocations of RAN_2 in parallel. This extension requires $2l$ multiplications carried out in 2 rounds.

Least Significant Bit Gate. The parties can compute the least significant bit of a secret shared value $[\![X]\!]$ of bounded size, i.e., $[\![X]\!] < 2^d$, using the LSB gate. Schoenmakers et al. [21] realize this protocol for Paillier encrypted values when there is sufficient "headroom" in the field, $p > 2^{d+\kappa+\log_2 n}$. This primitive can also be deployed in protocols based on secret sharing [16]. The implementation idea is that the parties jointly generate a random bit $[\![r_0]\!]$ and a uniformly random $(\kappa + d - 1)$-bit value $[\![r]\!]$ from which a random mask m is computed as $[\![m]\!] = 2[\![r]\!] + [\![r_0]\!]$. Then, $[\![c]\!] = [\![X]\!] + [\![m]\!]$ is computed and revealed. By the assumption on the size of p, we have that $c_0 = [\![X_0]\!] \oplus [\![m_0]\!]$. Thus, $[\![X_0]\!]$ can be obtained without cost as $[\![X_0]\!] = c_0 \oplus [\![m_0]\!] = c_0 + [\![m_0]\!] - 2c_0[\![m_0]\!]$. Overall, the complexity of this method is three multiplications in 2 rounds as $[\![r_0]\!]$ and $[\![r]\!]$ can be generated with two and one multiplication, respectively, in 2 rounds; the rest of the computation can be done without cost. From a security standpoint, c is statistically indistinguishable from random, preventing any adversary from learning information, even in the presence of corruption among all parties except one. Importantly, this is where the perfect security of our proposed protocol is lost, as the LSB gate is only statistically secure.

Table 1 summarizes the security guarantee and the number of multiplications and rounds required for one invocation of the main sub-protocols, where k represents the number of inputs to the prefix product.

Table 1. Complexity and security guarantees of the main sub-protocols used in this paper

Protocols	Rounds	Multiplications	Security
RAN_2	2	2	Perfect
Prefix product	3	$5k$	Perfect
LSB	2	3	Statistical

3   Secure MPC for Biased Coin Sampling

This section proposes a secure MPC protocol for a sampling technique that generates a biased coin, considering security against passive adversaries. Let b denote a coin with bias $0 < \alpha < 1$, and let α_B be the binary representation of α, which serves as a public input to the protocol. Recall that the protocol operates over the prime field $\mathbb{Z}_p$, where $l = \lceil \log_2 p \rceil$.

We use a sampling technique that involves three distinct steps:

1. Generate l uniformly random shared bits $[\![r]\!]_b$ using RAN_2.

2. Convert the comparison $[\![r]\!]_b > \alpha_B$ into the computation of a secret shared value $[\![X]\!]$. This is done so that $[\![X]\!]$ satisfies the condition $[\![X]\!] < 2^{\frac{l-1}{2}}$, and the least significant bit $[\![X_0]\!]$ of $[\![X]\!]$ corresponds to the result of $[\![r]\!]_b > \alpha_B$. The next subsection will provide details on this step.
3. Use the LSB gate to extract the least significant bit $[\![X_0]\!]$ of $[\![X]\!]$. Once the LSB gate is applied, $[\![\hat{b}]\!] = 1 - [\![X_0]\!]$ realizes the coin b with bias α. This realization exhibits a statistical difference of at most 2^{-l} when the shared bits $[\![r_{l-1}]\!], \ldots, [\![r_0]\!]$ are uniformly sampled from $\{0, 1\}$.

To successfully apply the LSB gate, the condition $p > 2^{d+\kappa+\log_2 n}$ must be met. Following an approach described by Reistad et al. [16], we can assume that $p > 2^{2(\kappa+\log_2 n)} = 2^{2d}$, ensuring the existence of at least $d = \kappa + \log_2 n$ bits of headroom within $\mathbb{Z}_p$. Having the assumptions $p > 2^{2d}$ and $l = \lceil \log_2 p \rceil$, we can set $l = 2d + 1$, an *odd* number. Necessarily $2^{2d} < p \leq 2^{2d+1}$ and Bertrand's postulate guarantees that such a prime p exists. This p is also an efficient choice as $p > 2^{2d+1}$ would increase the protocol's communication complexity.

The correctness of the sampling technique can be inferred from the previous discussion and the content presented in Sect. 3.1. The privacy of the sampling technique relies on the privacy provided by ABB and the underlying primitives employed in the method and the fact that the only revealed values occur in the primitives that have been considered in Sect. 2.2.

3.1 Detailed Computation of $[\![X]\!]$

Consider two l-bit values $[\![r]\!]_b$ and α_B. To determine if $[\![r]\!]_b > \alpha_B$, we compute a value $[\![x_i]\!]$ for each bit position i, ranging $\{0, 1, \ldots, l - 1\}$, as follows:

$$[\![x_i]\!] = \begin{cases} [\![r_i]\!](1 - \alpha_i) 2^{\sum_{j=i+1}^{l-1} \alpha_j \oplus [\![r_j]\!]} & \text{if } 0 \leq i < l - 1, \\ [\![r_i]\!](1 - \alpha_i) & \text{if } i = l - 1. \end{cases} \tag{1}$$

Next, we extract the least significant bit $[\![X_0]\!]$ from the sum:

$$[\![X]\!] = \sum_{i=0}^{l-1} [\![x_i]\!]. \tag{2}$$

The validity of this computation is confirmed by the following theorem, which is based extensively on Theorem 1 in the work of Reistad et al. [17].

Theorem 1. *Given l uniformly random shared bits $[\![r]\!]_b$ and a publicly known bit-decomposed value α_B, the expression $[\![\hat{b}]\!] = 1 - [\![X_0]\!]$ yields a shared coin with bias $0 < \alpha < 1$.*

Proof. The value $[\![x_i]\!]$ can take on either 0 or a unique power of 2, denoted as 2^k, where $k \in \{0, \ldots, l-2\}$. Specifically, $[\![x_i]\!] = 1$ if and only if $[\![r_i]\!]$ is set (equal to 1) and α_i is not set (equal to 0), while also satisfying $\sum_{j=i+1}^{l-1} \alpha_j \oplus [\![r_j]\!] = 0$

(this condition signifies that bit-position i is the most significant differing bit-position). Thus, when computing $[\![X]\!]$, it is guaranteed that at most one $[\![x_i]\!]$ value is odd, specifically when $[\![x_i]\!] = 1$, and this occurrence exists if and only if $[\![r]\!]_b > \alpha_B$. Finally, by evaluating $[\![\hat{b}]\!] = 1 - [\![X_0]\!]$, we obtain a shared coin with bias α. Notice that in the worst case, where all bits of $[\![r]\!]_b$ are set, and only one bit of α_B is not set (as $\alpha > 0$), the resulting value of $[\![X]\!]$ is equal to $2^{l-1} - 1$, thus it never needs a reduction modulo p.   □

To compute the values of $[\![x_i]\!]$, $2^{\sum_{j=i+1}^{l-1} \alpha_j \oplus [\![r_j]\!]}$ is performed for bit-positions $i = \{0, \ldots, l-2\}$. This can be done by rewriting the exponentiation of Eq. (1) as the following prefix product:

$$\prod_{j=i+1}^{l-1} (1 + (\alpha_j \oplus [\![r_j]\!])), \tag{3}$$

that can be computed as stated in Sect. 2.2. It is possible to convert from exponentiation to prefix product because all the terms $(1 + (\alpha_j \oplus [\![r_j]\!]))$ are invertible, as they are 1 or 2.

A remaining challenge to address is that, according to Eq.(2), $[\![X]\!] < 2^{l-1}$; however, for the LSB gate to be applied, we require $[\![X]\!] < 2^{\frac{l-1}{2}} = 2^d$. To overcome this issue, Reistad et al. [16,17] mention an extension of the technique described earlier for computing the values of $[\![x_i]\!]$. This extension involves considering pairs of bit-positions instead of individual bit-positions. Therefore, $[\![r_i]\!](1 - \alpha_i)$ is adjusted such that $r_i > \alpha_i$ for two bits at a time, and $\alpha_j \oplus [\![r_j]\!]$ outputs zero only when the corresponding pairs of bits are equal. By employing this extension, the number of $[\![x_i]\!]$ values is halved, thereby guaranteeing that $[\![X]\!] < 2^{\frac{l-1}{2}}$. We now present such an extension method as implemented by Zarei et al. [10].

Equation (4) is the two-bit variant of Eq. (1). It is specifically computed for bit-positions $i < l$ that are *even*, and the computation is based on the assumption that l is an *odd* value,

$$[\![x_i]\!] = [\![u_i]\!] 2^{\sum_{j=(i/2)+1}^{(l-1)/2} (\alpha_{2j} \oplus [\![r_{2j}]\!]) \vee (\alpha_{2j-1} \oplus [\![r_{2j-1}]\!])}, \tag{4}$$

where $[\![u_i]\!] = [\![r_i]\!](1 - \alpha_i)$ for $i = 0$, and $[\![u_i]\!] = [\![r_i]\!](1 - \alpha_i) + (1 + 2\alpha_i[\![r_i]\!] - [\![r_i]\!] - \alpha_i)[\![r_{i-1}]\!](1 - \alpha_{i-1})$ for $i = 2k, k \in \{1, 2, \ldots (l-1)/2\}$.

Furthermore, the exponentiation of Eq. (4) can be written as the following prefix product for implementation,

$$[\![v_i]\!] = \prod_{j=(i/2)+1}^{(l-1)/2} (1 + ([\![w_{2j}]\!] + [\![w_{2j-1}]\!] - [\![w_{2j}]\!][\![w_{2j-1}]\!])), \tag{5}$$

where $[\![w_{2j}]\!] = \alpha_{2j} \oplus [\![r_{2j}]\!]$, $[\![w_{2j-1}]\!] = \alpha_{2j-1} \oplus [\![r_{2j-1}]\!]$ and we set $v_{l-1} = 1$.

Finally, we can extract the least significant bit $[\![X_0]\!]$ of

$$[\![X]\!] = \sum_{i=0}^{(l-1)/2} [\![x_{2i}]\!] = \sum_{i=0}^{((l-1)/2)-1} [\![v_{2i}]\!][\![u_{2i}]\!] + v_{l-1}[\![u_{l-1}]\!]. \tag{6}$$

The complexity of evaluating $[\![X]\!]$ consists of performing $1.5(l-1)$ multiplications over three rounds. Additionally, there are $4(0.5(l-1))$ multiplications and two rounds required for the preparation of the prefix product in Eq. (5) as it takes on $(0.5(l-1))$ secret shared values. To elaborate further, $0.5(l-1)$ multiplications and one round are needed for computing $[\![u_i]\!]$'s that can be parallelized for implementation (note that the value α_i is a publicly known value, so $[\![r_i]\!](1-\alpha_i)$ can be considered without cost). These multiplications can be reused to calculate the terms of the prefix product in Eq. (5). Later, $0.5(l-1)$ multiplications and one round are required to mask the inputs of the prefix product in Eq. (5) (note that when this masking is done, prefixes can be computed without cost). Finally, $0.5(l-1)$ multiplications and one round are required for the $0.5(l-1)$ multiplications of Eq. (6) that can be parallelized for implementation (note that in Eq. (6) the value v_{l-1} is a publicly known value, so $v_{l-1}[\![u_{l-1}]\!]$ can be considered without cost).

Privacy of computing $[\![X]\!]$ is ensured as the secret sharing in use can only leak information when a value is revealed, and this occurs solely within the already considered sub-protocols. The result of this section can be summarized in the following theorem.

Theorem 2. *There exists a protocol to generate a shared biased coin using $O(l)$ secure multiplications in $O(1)$ rounds that is statistically secure against passive adversaries, when the arithmetic primitives have the same level of security.*

3.2   The Protocol Implementation and Complexity

We present Protocol 1 to implement the sampling technique and analyze the complexity involved. The complexity analysis of our proposed protocol involves measuring the number of multiplications required and the number of rounds needed to generate a biased coin. For a consistent basis of comparison, we note that $l = 2d + 1$ and present our complexity analysis in terms of d. This ensures a fair comparison of our protocol with that of Eriguchi et al. [11], discussed in Sect. 5.

In the first two rounds, we can generate $[\![r]\!]_b$ using $2l = 4d+2$ multiplications, in parallel with creating the random mask m (used for the LSB gate) using three multiplications and preparing the prefix product using $4(0.5(l-1)) = 4d$ multiplications. As discussed in Sect. 3.1, we can later compute $[\![X]\!]$ using $1.5(l-1) = 3d$ multiplications in 3 rounds. Lastly, $[\![X_0]\!]$ can be calculated at no additional cost using the LSB gate. In summary, generating one biased coin using this protocol requires $5.5l - 0.5 = 11d + 5$ multiplications and five rounds of computation.

4   Active Security Dissection

Security of our protocol against active adversaries can be achieved by using standard Verifiable Secret Sharing (VSS) schemes along with additional existing protocols. We now briefly expand on this claim.

Protocol 1 Multiparty Sampling a Coin with Bias α

Inputs: α_B and $l = \lceil \log_2 p \rceil$.
Output: A shared bit $[\![\hat{b}]\!]$, representing a coin with bias α.

1. Parties execute RAN_2 to generate $[\![r_i]\!]$ for $i \in \{0, 1, ..., l - 1\}$.
2. Parties compute $[\![u_i]\!] = [\![r_i]\!](1 - \alpha_i) + (1 + 2\alpha_i[\![r_i]\!] - [\![r_i]\!] - \alpha_i)[\![r_{i-1}]\!](1 - \alpha_{i-1})$ for $i = 2k, k \in \{1, \ldots ((l - 1)/2)\}$, where $[\![u_0]\!] = [\![r_0]\!](1 - \alpha_0)$.
3. Parties compute $[\![v_i]\!] =$
   $\prod_{j=(i/2)+1}^{(l-1)/2}(1 + ((\alpha_{2j} \oplus [\![r_{2j}]\!]) + (\alpha_{2j-1} \oplus [\![r_{2j-1}]\!]) - (\alpha_{2j} \oplus [\![r_{2j}]\!])(\alpha_{2j-1} \oplus [\![r_{2j-1}]\!])))$
   for $i = 2k, k \in \{0, 1, ..., ((l - 1)/2) - 1\}$, where $v_{l-1} = 1$.
4. Parties compute $[\![x_i]\!] = [\![u_{2i}]\!][\![v_{2i}]\!]$ for $i \in \{0, 1, ..., ((l - 1)/2) - 1\}$, where $[\![x_{(l-1)/2}]\!] = v_{l-1}[\![u_{l-1}]\!]$.
5. Parties locally compute $[\![X]\!] = \sum_{i=0}^{(l-1)/2}[\![x_i]\!]$.
6. Parties execute LSB gate with input $[\![X]\!]$ and obtain $[\![X_0]\!]$.
7. Parties locally compute $[\![\hat{b}]\!] = 1 - [\![X_0]\!]$ and output $[\![\hat{b}]\!]$.

In the context of active adversaries, all parties must prove that each computation step was performed correctly. If some dishonest parties withdraw from the process, others will be able to output (reveal) their shares and continue the rest of the computation. In particular, because any linear combination of shares is computed locally, each party must prove that it performed each multiplication correctly on its shares. Additional proofs include proofs that shares of a private value were distributed correctly among the parties (when the dealer is dishonest) and proofs of proper outputting (revealing) of a value from its shares. The above can be generally accomplished by employing a VSS scheme (see VSS approaches introduced by Cramer et al. [22] and Damgård et al. [23] that provide unconditional security, and VSS schemes presented by Pedersen [24] and Damgård et al. [25] that only guarantee computational security). In addition, if the parties are needed to generate input values of a specific form at any point of the computation, they would have to prove that their inputs are well-formed. Such proofs are required to implement the LSB gate, where the parties must show that their inputs to generate the random shared value $[\![r]\!]$ are of $(l + \kappa - 1)$ bit-length. This can be done through additional existing protocols discussed by Thorbek [26] with reasonable complexity. The technique employed is that any positive integer can be expressed as the sum of four squares; therefore, only eight multiplications are needed to show that an input is of both upper and lower bound.

5 Comparative Analysis

This section will begin with discussing Eriguchi et al.'s protocol [11], Π_d, and then compare it to our proposed protocol for sampling a biased coin.

Eriguchi et al. employ several primitives to construct Π_d. For $n \in \mathbb{N}$, $[n]$ denotes the set $\{z \in \mathbb{Z} : 1 \le z \le n\}$. A $\text{PRE}_\vee$ protocol computes d shares $[\![\vee_{k=1}^{j} a_k]\!]$, where $j, k \in [d]$, provided that the field size $p > 2d$. This protocol involves $17d$ multiplications and 7 rounds (including two rounds to generate random values using RAN_2). A XOR* protocol computes $[\![\oplus_{i\in[d]}b_i]\!]$ for $[\![b_i]\!] \in \{0,1\}$, using $5d$ multiplications in 1 round (including two rounds to generate random values). While there is a possibility of failure for these two protocols due to the use of RAN_2 and the inversion technique discussed in Sect. 2.2, the probability of these failures is typically on the order of p^{-1}, which can be considered negligible and disregarded (for more detailed information about these protocols, please refer to [20] and [27]).

Protocol 2 Multiparty Sampling a coin with bias α presented in [11]

Inputs: $s = ([\![s_{i1}]\!], \ldots, [\![s_{id}]\!])_{i\in[n]} \in (\{0,1\}^d)^n$.
Output: A shared bit $[\![\hat{b}]\!]$, representing a coin with bias α.

1. Each party $i \in [n]$ shares s_{il}, for $l \in [d]$.
2. Parties run XOR* with inputs $([\![s_{il}]\!])_{i\in[n]}$ and obtain $[\![r_l]\!] = [\![\oplus_{i\in[n]}s_{il}]\!]$ for $l \in [d]$.
3. Each party locally computes $[\![c_l]\!] = [\![r_l]\!] \oplus \alpha_l$ for $l \in [d]$.
4. Parties run PRE$_\vee$ with inputs $[\![c_l]\!]$, for $l \in [d]$ and obtain $[\![e_l]\!] = [\![\vee_{k=1}^{l}c_k]\!]$, for $l \in [d]$.
5. Parties locally compute $[\![f_l]\!] = [\![e_l]\!] - [\![e_{l-1}]\!]$ for $l \in [d]$, where $e_0 = 0$.
6. Parties compute $[\![f_l]\!][\![r_l]\!]$ for $l \in [d]$.
7. Parties compute $[\![\hat{b}]\!] = 1 - \sum_{l=1}^{d}[\![f_l]\!][\![r_l]\!]$ and output $[\![\hat{b}]\!]$.

Now, we provide a comprehensive explanation of how Π_d operates. Let b denote a coin with bias α, where $0 < \alpha < 1$, and for $l \in [d]$, let α_l be the l-th most significant bit in the binary representation of the public input α for Π_d. The protocol's idea involves generating d uniformly random bits $r_1, \ldots, r_d$, identifying the minimum index j where $r_j \neq \alpha_j$, and outputting $\hat{b} = 1 - r_j$. To implement this idea in the secure computation setting for $n > 2$ parties, we assume the field size $p > 2d$ and define a deterministic function $h_d^\alpha : (\{0,1\}^d)^n \to \mathbb{Z}_p$ defined by $h_d^\alpha(s) = 1 - r_j$, where $s = (s_{i1}, \ldots, s_{id})_{i\in[n]}$, $r_l = \oplus_{i\in[n]}s_{il}$ for $l \in [d]$, $r_{d+1} = 0$, and j is the smallest index such that $r_j \neq \alpha_j$ (if no such index exists, set $j = d+1$). Protocol 2 effectively implements Π_d, producing the shared outcome $\hat{b}$ that represents the coin b with bias α. When s is uniformly sampled from $(\{0,1\}^d)^n$, the statistical distance between b and $\hat{b}$ is at most 2^{-d}. As reported by Eriguchi et al., Π_d can be implemented using $(5n + 19)d$ multiplications and 11 rounds.

Table 2 compares the complexity of our solution to that of Eriguchi et al. [11].

Protocol 1 refers to our proposed protocol in this paper. Although this protocol admits statistical security, it allows a number of multiplications linear in d and a constant number of rounds for an unconditionally secure sampling of a biased coin. For this protocol, the field size must be sufficiently large to accommodate the large, random elements needed for applying the specified LSB gate. This necessitates an increase in our protocol's communication complexity. To account for this increase, we assumed that l (the number of bits representing the

Table 2. Complexity and security guarantees of biased coin protocols

	Rounds	Multiplications	Statistical difference	Security
Protocol 2	11	$(5n + 19)d$	2^{-d}	Perfect
Protocol 1	5	$11d + 5$	$2^{-(2d+1)}$	Statistical

field size) equals $2d + 1$. Based on this assumption, our protocol's complexity is reported in Table 2. Despite this increase, our protocol still requires fewer multiplications and rounds and produces a biased coin with a smaller statistical error than Protocol 2. It is important to note that, similar to our protocol, Protocol 2 requires the field size $p > 2d$ due to using the sub-protocol $\text{PRE}_\vee$. Both protocols may fail with a probability of the order p^{-1}; however, the failure probability can be arbitrarily small by selecting a large p. Our protocol can only demonstrate security against an active adversary when the parties prove their inputs for generating the random shared value $[\![r]\!]$ are of the specified bit-length while implementing the LSB gate. This can be efficiently obtained using additional protocols discussed in Sect. 4.

In line 2 of Protocol 2, the parties generate d uniformly random shared bits by XORing their local random bits, which costs $5nd$ multiplications. Another technique for achieving this, which we discussed in Sect. 2.2, is that the parties generate d uniformly random shared bits by d invocations of RAN_2 in parallel, incurring a cost of $2d$ multiplications. By employing this technique, Protocol 2 can be implemented using $20d$ multiplications and 10 rounds. Even with this change, our protocol still outperforms Protocol 2 in terms of the number of multiplications and rounds. This is shown in Table 3, where it is assumed that both protocols use RAN_2 to generate random shared bits.

Table 3. Complexity and security guarantees of biased coin protocols using RAN_2

	Rounds	Multiplications	Statistical difference	Security
Protocol 2	10	$20d$	2^{-d}	Perfect
Protocol 1	5	$11d + 5$	$2^{-(2d+1)}$	Statistical

6 Conclusion

In this paper, we have proposed an efficient, statistically secure protocol for secure MPC of a biased coin. The presented protocol uses $2d + 1$ unbiased coins to sample a biased coin, where d corresponds to the finite field size p with $\lceil \log_2 p \rceil = 2d + 1$. It is built on secure arithmetic operations in $\mathbb{Z}_p$ and can be realized using any linear secret sharing scheme.

The protocol takes a publicly known bias in bits as input and generates a sequence of random bits. It then compares this bias against the generated bits to extract the least significant bit from a secret shared value, culminating in a shared biased coin.

Our protocol offers complexity improvements over the method proposed by Eriguchi et al. [11], which involves $(5n + 19)d$ multiplications across 11 rounds. Furthermore, by adopting an alternative method to extract the least significant bit, our protocol eliminates the restrictive requirement that the field size p must be controlled, a limitation noted in the approach by Zarei et al. [10]. These enhancements in efficiency and broader applicability not only advance the field of MPC but also promise to improve the integration of MPC with differential privacy.

References

1. Radford, A., Wu, J., Child, R., Luan, D., Amodei, D., Sutskever, I.: Language models are unsupervised multitask learners. OpenAI blog 1 (2019). https://paperswithcode.com/paper/language-models-are-unsupervised-multitask
2. Devlin, J., Chang, M., Lee, K., and Toutanova, K.: BERT: pre-training of deep bidirectional transformers for language understanding. In Proceedings of the 2019 Conference of the North American Chapter of the Association for Computational Linguistics: Human Language Technologies, NAACL-HLT 2019, Minneapolis, MN, USA, June 2-7, 2019, Volume 1 (Long and Short Papers), Burstein, J., Doran, C., and Solorio. T. (eds.). Association for Computational Linguistics, pp. 4171–4186 (2019). https://doi.org/10.18653/v1/n19-1423
3. Dwork, C. and Roth, A.: The algorithmic foundations of differential privacy. Found. Trends® Theor. Comput. Sci. **9**, 211–407 (2014). https://doi.org/10.1561/0400000042
4. Dwork, C., Kenthapadi, K., McSherry, F., Mironov, I., Naor, M.: Our Data, Ourselves: privacy via distributed noise generation. In: Vaudenay, S. (eds) EUROCRYPT 2006. Lecture Notes in Computer Science, vol. 4004, pp. 486–503. Springer, Heidelberg (2006). https://doi.org/10.1007/11761679_29
5. Clement, C., Kamath, G., and Steinke, T.: The discrete Gaussian for differential privacy. J. Priv. Confidentiality **12**(1) (2022). https://doi.org/10.29012/jpc.784
6. The Differential Privacy Team at Google: Secure Noise Generation (2020). https://github.com/google/differential-privacy/blob/main/common_docs/Secure_Noise_Generation.pdf
7. Damgård, I., Nielsen, J.B.: Universally composable efficient multiparty computation from threshold homomorphic encryption. In: Boneh, D. (ed.) CRYPTO 2003. LNCS, vol. 2729, pp. 247–264. Springer, Heidelberg (2003). https://doi.org/10.1007/978-3-540-45146-4_15
8. Canetti, R.: Universally composable security: a new paradigm for cryptographic protocols. In: Proceedings 42nd IEEE Symposium on Foundations of Computer Science, pp. 136–145 (2001). https://doi.org/10.1109/SFCS.2001.959888
9. Champion, J., Shelat, A., Ullman, J.: Securely sampling biased coins with applications to differential privacy. In: Proceedings of the 2019 ACM SIGSAC Conference on Computer and Communications Security (CCS '19). Association for Computing Machinery, New York, NY, USA, 603–614 (2019). https://doi.org/10.1145/3319535.3354256

10. Zarei, A., et al. Computer Security. ESORICS 2023 International Workshops. ESORICS 2023. Lecture Notes in Computer Science, vol. 14398. Springer, Cham (2023). https://doi.org/10.1007/978-3-031-54204-6_19
11. Eriguchi, R., Ichikawa, A., Kunihiro, N., Nuida, K.: Efficient noise generation to achieve differential privacy with applications to secure multiparty computation. In: Borisov, N., Diaz, C. (eds) FC 2021. Lecture Notes in Computer Science, vol. 12674, pp. 271–290. Springer, Heidelberg (2021). https://doi.org/10.1007/978-3-662-64322-8_13
12. Bhaskar, R., Laxman, S., Smith, A., Thakurta, A.: Discovering frequent patterns in sensitive data. In: Proceedings of the 16th ACM SIGKDD International Conference on Knowledge Discovery and Data Mining, pp. 503–512 (2010). https://doi.org/10.1145/1835804.1835869
13. Shamir, A.: How to share a secret. Commun. ACM **22**, 612–613 (1979). https://doi.org/10.1145/359168.359176
14. Ben-Or, M., Goldwasser, S., Wigderson, A.: Completeness theorems for noncryptographic fault-tolerant distributed computations. In: Proceedings of the Twentieth Annual ACM Symposium on Theory of Computing, pp. 1–10. ACM Press, New York (1988). https://doi.org/10.1145/62212.62213
15. Reistad, T.I., Toft, T.: Secret Sharing Comparison by Transformation and Rotation. In: Desmedt, Y. (eds) ICITS 2007. Lecture Notes in Computer Science, vol. 4883, pp. 169-180. Springer, Heidelberg (2009). https://doi.org/10.1007/978-3-642-10230-1_14
16. Reistad, T.I., Toft, T.: Linear, constant-rounds bit-decomposition. In: Lee, D., Hong, S. (eds) ICISC 2009. Lecture Notes in Computer Science, vol. 5984, pp. 245–257. Springer, Heidelberg (2010). https://doi.org/10.1007/978-3-642-14423-3_17
17. Reistad, T.I.: Multiparty comparison-an improved multiparty protocol for comparison of secret-shared values. In: SCITEPRESS, vol. 1, pp. 325–330 (2009)
18. Toft, T.: Primitives and applications for multi-party computation. Ph.D. Dissertation, Aarhus University (2007). https://citeseerx.ist.psu.edu/document?repid=rep1&type=pdf&doi=66322abf90c2ff4357689b1a93acfd15dbfc4b03
19. Toft, T.: Constant-rounds, almost-linear bit-decomposition of secret shared values. In: Fischlin, M. (eds) CT-RSA 2009. Lecture Notes in Computer Science, vol. 5473, pp. 357–371. Springer, Heidelberg (2009). https://doi.org/10.1007/978-3-642-00862-7_24
20. Damgård, I., Fitzi, M., Kiltz, E., Nielsen, J.B., Toft, T.: Unconditionally secure constant-rounds multi-party computation for equality, comparison, bits and exponentiation. In: Halevi, S., Rabin, T. (eds) TCC 2006. Lecture Notes in Computer Science, vol. 3876, pp. 285–304. Springer, Heidelberg (2006). https://doi.org/10.1007/11681878_15
21. Schoenmakers, B., Tuyls, P.: Efficient binary conversion for Paillier encrypted values. In: Vaudenay, S. (eds) EUROCRYPT 2006, LNCS, vol. 4004, pp. 522–537. Springer, Heidelberg (2006). https://doi.org/10.1007/11761679_31
22. Cramer, R., Damgård, I., Maurer, U.: General secure multi-party computation from any linear secret-sharing scheme. In: Preneel, B. (eds) Advances in Cryptology - EUROCRYPT 2000. Lecture Notes in Computer Science, vol. 1807, pp. 316–334. Springer, Heidelberg (2000). https://doi.org/10.1007/3-540-45539-6_22
23. Damgård, I., Ishai, Y., Krøigaard, M.: Perfectly secure multiparty computation and the computational overhead of cryptography. In: Gilbert, H. (eds) Advances in Cryptology - EUROCRYPT 2010. Lecture Notes in Computer Science, vol.

6110, pp. 445–465. Springer, Heidelberg (2010). https://doi.org/10.1007/978-3-642-13190-5_23

24. Pedersen, T.P.: Non-interactive and information-theoretic secure verifiable secret sharing. In: Feigenbaum, J. (eds) Advances in Cryptology - CRYPTO '91. Lecture Notes in Computer Science, vol. 576, pp. 129–140. Springer, Heidelberg (1992). https://doi.org/10.1007/3-540-46766-1_9

25. Damgård, I., Ishai, Y., Krøigaard, M., Nielsen, J.B., Smith, A.: Scalable multiparty computation with nearly optimal work and resilience. In: Wagner, D. (eds) Advances in Cryptology - CRYPTO 2008. Lecture Notes in Computer Science, vol. 5157, pp. 241–261. Springer, Berlin, Heidelberg (2008). https://doi.org/10.1007/978-3-540-85174-5_14

26. Thorbek, R.: Linear integer secret sharing. Ph.D. Dissertation, Aarhus University (2009)

27. Nishide, T., Ohta, K.: Multiparty computation for interval, equality, and comparison without bit-decomposition protocol. In: Okamoto, T., Wang, X. (eds) PKC 2007. LNCS, vol. 4450, pp. 343–360. Springer, Heidelberg (2007). https://doi.org/10.1007/978-3-540-71677-8_23

28. Eriguchi, R., Ichikawa, A., Kunihiro, N., Nuida, K.: Efficient noise generation protocols for differentially private multiparty computation. IEEE Trans. Dependable Secure Comput. 1, 1–16 (2022). https://doi.org/10.1109/TDSC.2022.3227568

A Formal Approach for Modelling and Analysing Surgical Procedures

Ioana Sandu[1]([✉])(iD), Rita Borgo[1](iD), Prokar Dasgupta[2](iD), Ramesh Thurairaja[3], and Luca Viganó[1](iD)

[1] Department of Informatics, King's College London, London, UK
{ioana.sandu,rita.borgo,luca.vigano}@kcl.ac.uk
[2] Peter Gorer Department of Immunobiology, King's College London, London, UK
prokar.dasgupta@kcl.ac.uk
[3] Urology Department, Guy's and St Thomas' Hospital NHS Foundation Trust, London, UK
Ramesh.Thurairaja@gstt.nhs.uk

Abstract. Surgical procedures are often not "standardised" (i.e., defined in a unique and unambiguous way), but rather exist as implicit knowledge in the minds of the surgeon and the surgical team. This reliance extends to pre-surgery planning and effective communication during the procedure. We introduce a novel approach for the formal and automated analysis of surgical procedures, leveraging established techniques developed for the analysis of security ceremonies. Our approach allows us to model as mutations the variants of a procedure and the mistakes that members of the surgical team might make, and to automatically identify violations of the intended properties of a procedure.

Keywords: Formal Methods · Mistakes in Surgical Procedures · Variants of Surgical Procedures · Security Ceremonies

1 Introduction

Context and Motivation. This paper is the result of a collaboration between computer scientists and clinician scientists, which commenced with the live observation of a robot-assisted prostatectomy and cystectomy, leading to in-depth discussions on the actual execution of surgical procedures. These emphasised that much of a surgical procedure is often in the heads of the surgeon and of the members of the surgical team. This reliance on internalised knowledge hinges on two critical activities: (1) comprehensive pre-surgery discussions between the team members, (2) effective communication throughout the procedure. Recognising the potential for errors in both activities, which could jeopardise patient safety, *surgical process models (SPMs)* have been proposed to represent surgical procedures. These models offer "simplified, formal, or semi-formal representations of a network of surgery-related activities" [7].[1] SPMs draw upon concepts from workflow management and computer science, and (often) provide a representation of a surgical

---

[1] See [3,5,14,17] for SPMs for robot-assisted prostatectomies and similar procedures.

© The Author(s), under exclusive license to Springer Nature Switzerland AG 2025
F. Martinelli and R. Rios (Eds.): STM 2024, LNCS 15235, pp. 82–93, 2025.
https://doi.org/10.1007/978-3-031-76371-7_6

procedure that can be communicated to team members as well as to other surgeons so they may follow the same steps. However, even in the case of more formal SPMs, little to no attention has been devoted to using SPMs to reason about procedures, particularly in the context of (1) and (2).

Contributions. We propose a different approach but one still anchored in computer science and, more specifically, cybersecurity: we formally model and reason about surgical procedures by representing them as security ceremonies. Modelling a surgical procedure as a security ceremony brings some important advantages. It provides conceptual clarity and allows one to represent the procedure as a message sequence chart that can be published and shared with others. It also provides a structured framework for reasoning as it enables us to adapt to surgical procedures established methodologies and automated approaches developed for the formal analysis of security ceremonies and their properties.

A *security protocol* is essentially a communication protocol (an agreed sequence of actions performed by agents communicating to accomplish some mutually desirable goals) that makes use of cryptographic techniques, allowing the communicating agents to satisfy security properties, such as authentication, or confidentiality or integrity of data. A *security ceremony* expands a security protocol to include human nodes alongside computer nodes, with communication links that comprise user interfaces, human-to-computer and human-to-human communication, and transfers of physical objects that carry data [13]. Hence, a ceremony's analysis should include the mistakes that human agents might make when they execute their tasks. Modelling a surgical procedure as a security ceremony thus allows us to consider *mutations* of the ceremony/procedure that formalise possible mistakes made by members of the surgical team. For instance, we formally model that the actions of the surgical team should not cause internal bleeding or endanger the patient, and our approach allows us to capture violations of this property, e.g., situations where a surgeon performs an internal incision without the assistant applying clips to prevent bleeding.[2] To automatically identify that such mistakes violate the intended properties of the procedure, we adapt and extend the mutation-based analysis approach proposed in [9] (but we use a different tool, UPPAAL [4], as it provides greater visual simplicity). Also, mutations provide the means for researchers and surgeons to explore variants of the procedure (e.g., alterations in the order of actions) and check if they lead to property violations without having to perform the variant on a live patient.

To illustrate our approach, we consider two stages of a laparoscopic prostatectomy procedure that is described informally in [3], which provides one of the most comprehensive descriptions of a laparoscopic prostatectomy. However, the description in [3], as is standard in such papers, is purely textual (with a few anatomical illustrations) and thus informal and prone to misunderstandings, so providing a formal model as we do is already a valuable contribution.[3]

---

[2] We focus on mistakes by human agents, but mistakes by robotic agents could be considered similarly.

[3] Note that [3] is 20+ years old and some specifics of the standard laparoscopic prostatectomy procedure might have changed in the meantime (cf. the newer papers discussed in Sect. 2) but adapting our models and analysis accordingly would be quite easy.

Structure. In Sect. 2, we discuss background and related work. We present our formal model in Sect. 3, the mutations in Sect. 5, and the formal analysis in Sect. 5. We draw conclusions in Sect. 6. Full details of our models and analyses are provided in [8].

2   Background and Related Work

Robotic-assisted surgery (RAS) has transformed the conventional operating room by introducing changes that include increased spatial requirements due to equipment and the physical separation of console surgeons from patients and team. In contrast to traditional arrangements [10], the configuration of RAS may hinder interpersonal cues and lead to potential miscommunication. Our approach proposes to reason about surgical procedures by conceptualising them as security ceremonies, which offer an explicit representation of human agents and their communications with other agents (human or not) [2,9]. This perspective enables us to systematically incorporate and reason about human mistakes in the context of RAS or surgical procedures of any kind. We could similarly model robotic agents in RAS and other features of such procedures. This is important as RAS encompasses the patient, surgery type, surgical goals, tasks contributing to those goals, patient-related factors, and situational factors. The integration of new technologies into the operating room has the potential to significantly alter the prerequisites for effective teamwork, procedural workflows, and individual skills [11]. The distinctive setup of RAS introduces new challenges in maintaining situational awareness, team coordination, and information exchange [15]. Hence, effective communication is crucial for maintaining a surgeon's situational awareness, especially when operating from the console [16]. Communication, a recognised source of disruption in surgeries, has been undergoing fundamental changes in RAS due to the relocation of the surgeon from the operating table, and the impact of workflow disruptions/interruptions is explored in, e.g., [15,16]. Specific verbal/non-verbal cues are crucial for team coordination [1], and studies have delved into the influence of anticipation and teamwork in RAS [10].

Excellent methods for conventional laparoscopic radical prostatectomy have been described in, e.g., [3,5,17], but there is currently no standardised surgical technique for robot-assisted radical prostatectomy.

In this paper, we demonstrate the efficacy of our approach on a simple stage of the procedure, the cutting stage, and on a more intricate stage, the dissection of the lateral surfaces of the prostate. The latter stage is pivotal because preserving the neurovascular bundles is paramount for ensuring a successful surgical outcome for patients who aim to maintain postoperative potency. Failure to preserve these bundles could significantly impact such patients' recovery. Various approaches to nerve-sparing prostatectomies are discussed in [6]. Denonvilliers' fascia is a crucial structure covering the posterior surface of the prostate and separating it from the rectum. It plays a vital role in the confinement of cancer within the prostate and facilitating an operation without damaging the nerves responsible for erectile function and continence, while ensuring the removal of all neoplastic tissue [12]. Hence, this stage not only demonstrates the close collaboration between surgeon and assistant but also allows us to reason about one of the key factors contributing to a successful outcome and recovery.

3   Formal Model

In a surgical procedure, multiple agents collaborate through a series of orchestrated actions and message exchanges to execute their tasks seamlessly as a cohesive team. We thus model a surgical procedure as a sequence of actions and messages exchanged so that other actions can occur. As a concrete example, we provide a formal model (and security analysis) of two stages of a laparoscopic prostatectomy procedure that is described informally in [3]. The *message sequence chart (MSC)* in Fig. 1a shows the cutting stage of this procedure, where three agents, a surgeon S, an assistant A and a nurse N, collaborate to carry out an internal incision on a patient.[4] Figure 1b shows the MSC that we have drawn for the *lateral dissection stage* (the dissection of the lateral surfaces of the prostate), where VD, DF, SV and NVB abbreviate Denonvilliers' Fascia, vasa deferentes, seminal vesicles and neurovascular bundles, respectively.

In an MSC, each agent is defined as a process characterised by a series of surgical actions (the boxes in each agent's vertical timeline) and messages they send (the horizontal arrows), confirming that they have carried out an action or requested other agents to carry out an action. As is standard for security protocols/ceremonies (e.g., [9,13]), we define the *algebra of messages* as $T_\Sigma(\mathcal{V})$. The signature Σ contains possibly disjoint sets of constants (e.g., representing agent names and other publicly known values) and $\mathcal{V}$ is a countably infinite set of variables. Σ can easily be extended to include function symbols to formalise symmetric and asymmetric decryption and other cryptographic operators. Given the set M of all messages that can be built according to the algebra, we define for each agent Ag the sets M_{Ag}^s and M_{Ag}^r of messages Ag can send and receive. Here, we only consider messages that can be defined as constants, as that is what our case study requires. For instance, for S in both stages of the procedure we define $M_S^s = \{$ *clips_requested, cut_done, VD_and_SV_pulled, pedicle_dissected, PFS_entered, visceral_fascia_incised, DF_incised* $\}$. Our approach can accommodate more complex messages, e.g., that contain random numbers and are encrypted. Moreover, we consider only honest agents who behave according to what the surgical procedure expects, but below we will extend this to consider mistakes by agents (and we could also consider dishonest agents who can do anything they want as is standard in formal analysis of security ceremonies).

In formal analysis of security ceremonies, and thus in our approach, agents are formalised as processes that represent the vertical lines in an MSC and that are often called role-scripts. A *role-script* is a sequence of events $e \in T_{\Sigma \cup RoleActions}$, where $RoleActions = \{Snd, Rcv, s\_action, Start\}$ is a set of action names with their respective arity. For example, the role-script of the cutting stage of S is

---

[4] Note that this cutting stage is quite general and could be applied also to other surgical procedures and not just to a prostatectomy.

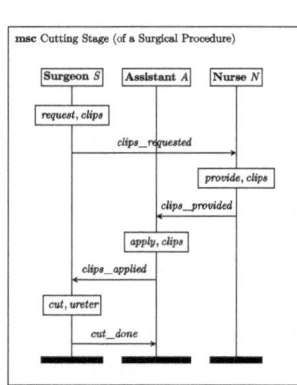

(a) MSC of the Cutting Stage

$RoleScript_S =$
$[Start(S, K_S^0)$
$s\_action(S, request, clips)$
$Snd(S, N, clips\_requested)$
$Rcv(A, S, clips\_applied)$
$s\_action(S, cut, ureter)$
$Snd(S, A, cut\_done)]$

(b) Role-script of S

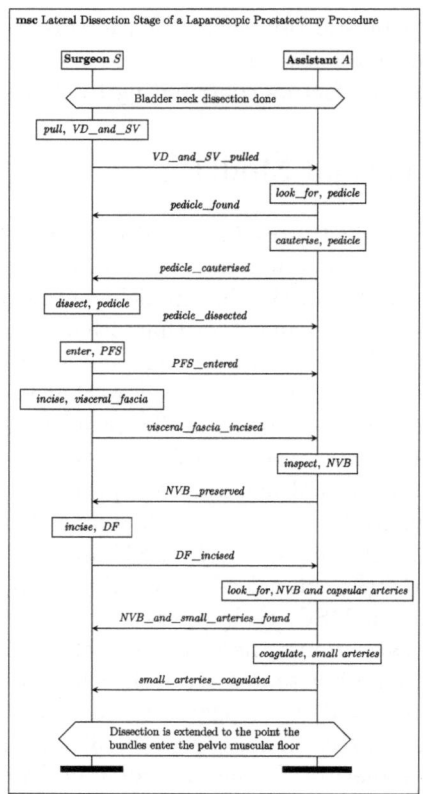

(c) MSC of the Lateral Dissection Stage

Fig. 1. MSCs of Two Stages of a Laparoscopic Prostatectomy Procedure and Role-Script of the Surgeon in the Cutting Stage

shown in Fig. 1b. $Start(Ag, K_{Ag}^0)$ is the first event of a role-script and it takes place once, where K_{Ag}^0 is the *initial knowledge* of agent Ag at the beginning of the process (typically, it contains the names of the other agents and the constant messages Ag will send). Ag's *knowledge* increases monotonically as Ag receives messages. Snd and Rcv events are of the form $Snd(Ag_s, Ag_r, m)$ and $Rcv(Ag_s, Ag_r, m)$, where Ag_s is the sender of the message, Ag_r is the receiver and m is the message that is being sent or received. In our model, the messages have only one recipient to indicate who the information concerns; this is primarily a modeling choice for clarity, but it can be adjusted so that multiple agents receive the same message. Moreover, we focus on secure communication channels between agents, but our approach extends to various types of channels (e.g., authenticated or insecure ones, which can be attacked by a dishonest agent).

Surgical action events represent the actions each agent performs to ensure the progression of the surgery and are defined as $s\_action(Ag, a_{Ag}, O)$, agent Ag performs action a_{Ag} on or with object O. For the two stages we are considering,

$a_S \in \{request, cut, pull, dissect, enter, incise\}$, $a_A \in \{apply, look\_for, cauterise, inspect, coagulate\}$ and $a_N \in \{provide\}$.[5] The objects are $O = \{clips, ureter, VD\_and\_SV, pedicle, PFS, visceral\_fascia, NVB, DF, capsular arteries, small arteries\}$, which are significant as the same action may be performed multiple times for different objects (e.g., one can request clips or a scalpel), or an action could be performed with the same object but at different stages of the procedure (e.g., clips can be requested during cutting or suturing).

Our approach is based on an *execution model* that is defined by a *multi-set rewriting system* like in many security protocol/ceremony analysis approaches and tools. A *state* is a multiset of facts that model resources, including the information that agents know and exchange. Formally, the state $S^i_{Ag} = \{i; K^i_{Ag}\}$ of agent Ag is characterised by the state number i and the knowledge $K^i_{Ag} = K^0_{Ag} \cup \{messages\ received\ by\ Ag\}$ that Ag possesses at i, and $S^i = \{i; \{K^i_{Ag} \mid Ag\ \text{is an agent}\}\}$ represents the state of all agents at that point in the execution.

A *trace* is a finite sequence of multisets of role-actions and is generated by the application of *state transition rules* of the form

$$prem \xrightarrow{\textit{finite sequence of role−actions and internal checks}} conc$$

which is applicable when the current state matches the premise *prem* and the internal checks on the messages received are satisfied. These checks are typically not displayed in a role-script but only act as guards. The rule's application produces the conclusion *conc* (a new state) and records the instantiations of role-actions in the trace. For instance, for the cutting stage,

$$\{1;\ K^1_S, K^1_A, K^1_N\} \xrightarrow{\textit{Rcv}(S, N, ?X),\ \textit{Check}(?X = clips\_requested),}$$
$$\xrightarrow{\textit{s\_action}(N, provide, clips),\ \textit{Snd}(N, A, clips\_provided)} \{2;\ K^2_S, K^2_A, K^2_N\}$$

represents a transition from state S^1 to state S^2, where we split the arrow for readability and where we are numbering the states from 0 to 5 assuming that the cutting stage is the initial stage of the procedure (if it is not the initial stage, then the states will be numbered differently but still consecutively). Agent N receives a message $?X$ from S, checks the contents of that message, performs a surgical action and sends a confirmation message to A. We write that agent Ag receives $?X$ to allow Ag to check, via $Check(?X = m)$, that this is indeed the message m that Ag was expecting. This check will become useful later as it will enable us to consider mistakes that agents could make, such as changing the contents of the message or sending the wrong message.[6]

---

[5] We could represent cuts and incisions by means of a single action but we prefer to consider two distinct actions *cut* and *incise* to distinguish between actions that might use different instruments (e.g., scissors or scalpels).

[6] In security ceremonies, there typically is a ? also in front of the sender's name in a *Rcv* event. This allows one to consider an attacker that is claiming to be the sender. Here, we avoid doing so given that we are not yet explicitly considering an attacker.

Nurse N can carry out only the transition above. Surgeon S's can carry out

$$\{0;\ K_S^0, K_A^0, K_N^0\} \xrightarrow{\textit{Start, s\_action(S, request, clips), Snd(S, N, clips\_requested)}} \{1;\ K_S^1, K_A^1, K_N^1\}$$

$$\{3;\ K_S^3, K_A^3, K_N^3\} \xrightarrow{\substack{\textit{Rcv(A, S, ?X), Check(?X = clips\_applied),} \\ \textit{s\_action(S, cut, ureter), Snd(S, A, cut\_done)}}} \{4;\ K_S^4, K_A^4, K_N^4\}$$

and assistant A can carry out

$$\{2;\ K_S^2, K_A^2, K_N^2\} \xrightarrow{\substack{\textit{Rcv(N, A, ?X), Check(?X = clips\_provided),} \\ \textit{s\_action(A, apply, clips), Snd(A, S, clips\_applied)}}} \{3;\ K_S^3, K_A^3, K_N^3\}$$

$$\{4;\ K_S^4, K_A^4, K_N^4\} \xrightarrow{\textit{Rcv(S, A, ?X), Check(?X = cut\_done)}} \{5;\ K_S^5, K_A^5, K_N^5\}$$

The expected sequence of actions is then given by the rules applied in the same order as their state numbers, mirroring the MSC. It captures the unaltered process, when all agents execute their tasks precisely as expected. No errors are occurring, but rather every action unfolds in a seamless sequence, with agents patiently awaiting messages from their predecessors before proceeding.

The transition rules for the lateral dissection stage are similar, e.g.

$$\{i;\ K_S^i, K_A^i, K_N^i\} \xrightarrow{\substack{\textit{Rcv(A, S, ?X), Check(?X = NVB\_preserved),} \\ \textit{s\_action(S, incise, DF), Snd(S, A, DF\_incised)}}} \{i+1;\ K_S^{i+1}, K_A^{i+1}, K_N^{i+1}\}$$

represents the incision of the DF between states i and $i+1$ (we use i to indicate the state as it occurs in the full procedure when all stages are considered).

4   Mutations

When engaged in a surgical procedure (and, in general, in a security ceremony), humans might make mistakes because of various reasons, such as communication errors, distraction, inexperience, stress, etc. These mistakes alter the process flow and create deviations of the original ceremony specification that may impact the security of the ceremony or, in a surgical procedure, the safety of the patient.

We adapt and extend to surgical procedures the approach of [9], which allows security analysts to model mistakes by human agents as *mutations* of the behaviour that the ceremony originally specified for such agents.[7] Mutations thus create alternative formal specifications of the original ceremony, which we can then formally analyse to see if they lead to violations of the intended properties (and thus endanger patient safety). Studying these mutations is also interesting as they might reveal alternative ways to carry out the procedure that do not endanger the patient and are, possibly, faster or more efficient. In this paper, we focus on formal analysis, dis-/proving properties of a procedure, but in the future we plan to also carry out a cost analysis of the different secure alternatives.

Since mutations allow humans to do things that were not foreseen in the original procedure, we formalise them by introducing new transition rules that are themselves mutations of the original ones. For surgical procedures, we focus on two mutations, skip and replace (but more could be considered). The *skip mutation* enables us to formalise an agent skipping some actions that the surgical procedure expects them to carry out. For instance, for the cutting stage, the case in which A does not apply the clips but nonetheless sends a message confirming task completion is formalised by the mutated rule

$$\{2;\ K_S^2, K_A^2, K_N^2\} \xrightarrow[\text{$s\_action(A,\ apply,\ clips)$}, \ Snd(A,\ S,\ clips\_applied)]{Rcv(N,\ A,\ ?X),\ Check(?X\ =\ clips\_provided),} \{3;\ K_S^3, K_A^3, K_N^3\}$$

and the case in which A applies the clips but does not send a confirmation message is formalised by the mutated rule

$$\{2;\ K_S^2, K_A^2, K_N^2\} \xrightarrow[s\_action(A,\ apply,\ clips),\ \text{$Snd(A,\ S,\ clips\_applied)$}]{Rcv(N,\ A,\ ?X),\ Check(?X\ =\ clips\_provided),} \{3;\ K_S^3, K_A^3, K_N^3\}$$

$$\{2;\ K_S^2, K_A^2, K_N^2\} \xrightarrow[s\_action(A,\ apply,\ clips),\ \text{$Snd(A,\ S,\ clips\_applied)$}]{Rcv(N,\ A,\ ?X),\ Check(?X\ =\ clips\_provided),} \{3;\ K_S^3, K_A^3, K_N^3\}$$

In the *replace mutation*, an agent performs their actions as expected but replaces a message with another one. For instance, in case of complex messages consisting of different components, an agent could send just part of the message by mistake, as considered in the mutations in [9]. For our case study, where messages are simple, we introduce the novel (w.r.t. [9]) concept of *negative message*, which we write as "not_m", e.g., not_clips_applied, and we extend accordingly the sets of messages agents can send or receive. This allows agents in the cutting stage of our case study to execute mutated rules such as:

---

[7] These mutations refer to deviations from the expected sequence of actions in a process or procedure, not to the mutations found in other fields such as molecular biology or genetic mutations. We use the mutation names given in [9] but other names have been proposed for similar mutations in different disciplines, e.g., in biology.

$$\{2; \ K_S^2, K_A^2, K_N^2\} \xrightarrow[\quad s\_action(A, \ apply, \ clips), \ Snd(A, \ S, \ not\_clips\_applied) \quad]{\quad Rcv(N, \ A, \ ?X), \ Check(?X = clips\_provided), \quad} \{3; \ K_S^3, K_A^3, K_N^3\}$$

Each action in a surgical procedure has a purpose and altering even a single action will cause some sort of propagation of the mistake for the next agents, which could impact patient safety. When a mutation happens, the other agents will not be able to carry out their actions unless their rules are mutated as well. For instance, if A does not apply the clips, then S will not cut and execution will deadlock. Our aim is for a procedure not to deadlock during execution but rather to be executed completely so that we can check whether the intended properties are satisfied even in presence of a mistake. To ensure that we only have executable traces, every mutation is matched via a *matching mutation* and propagated through a trace. A matching mutation for a skip mutation depends on the ability of an agent to perform their action given that the previous agent has skipped theirs. For example, if N skips their action to provide the clips

$$\{1; \ K_S^1, K_A^1, K_N^1\} \xrightarrow[\quad \text{s\_action}(N, \ \text{provide}, \ \text{clips}), \ Snd(N, \ A, \ clips\_provided) \quad]{\quad Rcv(S, \ N, \ ?X), \ Check(?X = clips\_requested), \quad} \{2; \ K_S^2, K_A^2, K_N^2\}$$

the matching mutation formalises that the action A was going to perform, apply clips, is skipped as it is impossible for A to apply clips unless N provides them

Other mutations are matched and propagated similarly (cf. [8] for details).

$$\{2; \ K_S^2, K_A^2, K_N^2\} \xrightarrow[\quad \text{s\_action}(A, \ \text{apply}, \ \text{clips}), \ Snd(A, \ S, \ clips\_applied) \quad]{\quad Rcv(N, \ A, \ ?X), \ Check(?X = clips\_provided), \quad} \{3; \ K_S^3, K_A^3, K_N^3\}$$

5   Formal and Automated Analysis

Surgical procedures should first and foremost guarantee patient safety. Hence, everything that might endanger it should be avoided and formally specified as a *property* to be satisfied. This way, we can formally analyse it and either prove it to hold or, if not, produce a trace that shows the sequence of actions violating the property (if the tool terminates). We formalise properties using a linear temporal logic, which allows us to specify that if an event occurs now, then certain other events must have occurred in the past. For instance, the patient should not bleed out due to a negligent incision, i.e., we require that in all traces, if S carries out a cut at some time instant, then there must exist previous time instants, ordered temporally, in which the clips have been requested, provided and applied:

Property 1 (Clip-before-cutting). For all traces,

$$s\_action(S, cut) @l \implies s\_action(S, \ request, \ clips) @i$$
$$\& \ s\_action(N, \ provide, \ clips) @j$$
$$\& \ s\_action(A, \ apply, \ clips) @k$$
$$\& \ i < j < k < l$$

Property 1, which is a general and quite obvious property of any surgical procedure but is also explicitly inspired by the informal discussion in [3], establishes a fundamental sequence of actions that occur in any procedure that includes a cutting stage (cf. Fig. 1a). The following three properties consider, instead, the lateral dissection stage (cf. Fig. 1b) and are again inspired by [3] as well as by the more recent [6,12]. Property 2 pertains to the dissection of the pedicle, Property 3 checks whether the incision of DF has been performed, Property 4 expresses that preserving the NVB is crucial for potency recovery [6].

Property 2 (Dissection of the pedicle). For all traces,

$$s\_action(S, dissect, pedicle) @l \implies s\_action(S, \ pull, \ VD\_and\_SV) @i$$
$$\& \ s\_action(A, \ look\_for, \ pedicle) @j$$
$$\& \ s\_action(A, \ cauterise, \ pedicle) @k$$
$$\& \ i < j < k < l$$

Property 3 (Incision of the Denonvilliers' Fascia). For all traces,

$$s\_action(S, incise, DF) @l \implies s\_action(S, enter, PFS) @i$$
$$\& \ s\_action(S, incise, visceral\_fascia) @j$$
$$\& \ s\_action(A, inspect, NVB) @k$$
$$\& \ i < j < k < l$$

Property 4 (Check if the nerves are preserved). For all traces,

$$s\_action(A, inspect, NVB) @l \implies s\_action(A, \ cauterise, \ pedicle) @i$$
$$\& \ s\_action(S, \ enter, \ PFS) @j$$
$$\& \ s\_action(S, \ incise, \ visceral\_fascia) @k$$
$$\& \ i < j < k < l$$

We automatically analysed these properties using UPPAAL [4]; see [8] for details on the modelling and analysis using UPPAAL and on the attack traces it outputs (and for other properties that could be considered). UPPAAL confirmed that Property 1 holds true across all traces when mutations are deactivated, but intriguing violations caused by agent mistakes become discernible upon enabling mutations (and matching mutations), e.g., the surgeon can execute a cut without the application of clips if mistakes occur due to miscommunication or negligence. Similarly, Property 2 suffers from an attack in which S dissects the pedicle without VD and SV being pulled and without the pedicle being cauterised. This result could indicate several possibilities: the pedicle might have been visible without the need to pull VD and SV and there might have been no bleeding requiring cauterisation. Although the patient's safety may not be compromised, this still represents a violation of the expected sequence of actions. Thus, it is crucial for a surgeon to interpret the results, as this attack could represent a legitimate shortcut in the procedure. Our approach thus allows clinicians to consider variants of the procedure and reason about them.

Property 3 is violated when S makes an incision of the DF without incising the visceral fascia and without A inspecting the NVB. Property 4 is violated when the NVB have been inspected without entering the pericapsular fatty space and without incising the visceral fascia. It is important to ensure that Property 4 holds because of the presence of numerous NVB fibers between the posterior and intermediate layers of the DF, which makes dissection in this area hazardous for erectile nerves and should be avoided in nerve-sparing procedures [12].

6   Concluding Remarks

We view this paper as the first step towards the full-fledged analysis of surgical procedures. We plan to encompass a complete laparoscopic prostatectomy (and other procedures) by modelling and analysing all stages holistically rather than independently. We will explore methodologies for performing a prostatectomy, involving an expanded set of agents, both human and robotic, capable of executing a broader range of actions and transmitting additional messages. We will also consider telesurgeries, emphasising the importance of cryptography for secure communication between agents in the presence of possible attackers.

We also plan to consider other mutations, such as a mutation that "negates" an action by undoing it, which would, e.g., capture the mistake that occurs when A applies the clips but then removes them before S cuts, resulting in bleeding.

Although we could not discuss them here, our approach also allows one to explore variants of a procedure not only to identify attacks but also to study, say, an order of the actions different from that in the minds of the surgical team. Our objective is to offer the most suitable surgical approach for each individual, tailoring the procedure to meet specific patient needs and conditions.

References

1. Catchpole, K., et al.: Human factors in robotic assisted surgery: lessons from studies 'in the Wild'. Appl. Ergon. **78**, 270–276 (2019)
2. Curzon, P., Rukšėnas, R.: Modelling the user. In: The Handbook of Formal Methods in Human-Computer Interaction, pp. 211–245. Springer (2017)
3. Guillonneau, B., Vallancien, G.: Laparoscopic radical prostatectomy: the montsouris technique. J. Urol. **163**(6), 1643–1649 (2000)
4. Larsen, K.G., Pettersson, P., Yi, W.: UPPAAL (1995). https://uppaal.org
5. Martini, A., et al.: Contemporary techniques of prostate dissection for robot-assisted prostatectomy. Eur. Urol. **78**(4), 583–591 (2020)

6. Moschovas, M.C., Patel, V.: Neurovascular bundle preservation in robotic-assisted radical prostatectomy: how i do it after 15.000 cases. Int. Braz. J. Urol. **48** (2022)
7. Neumuth, T.: Surgical process modeling. Innov. Surg. Sci. **2**(3), 123–137 (2017)
8. Sandu, I., Borgo, R., Dasgupta, P., Thurairaja, R., Viganò, L.: A Formal Approach For Modelling And Analysing Surgical Procedures (Extended Version) (2024). https://arxiv.org/abs/2408.05001
9. Sempreboni, D., Viganò, L.: A mutation-based approach for the formal and automated analysis of security ceremonies. J. Comput. Secur. **31**(4), 293–364 (2023)
10. Sexton, K., et al.: Anticipation, teamwork and cognitive load: chasing efficiency during robot-assisted surgery. BMJ Qual. Saf. **27**(2), 148–154 (2018)
11. Souders, C.P., et al.: Flow disruptions in robotic-assisted abdominal sacrocolpopexy: does robotic surgery introduce unforeseen challenges for gynecologic surgeons? Int. Urogynecol. J. **30**(12), 2177–2182 (2019)
12. Tzelves, L., Protogerou, V., Varkarakis, I.: Denonvilliers' fascia: the prostate border to the outside world. Cancers **14**(3) (2022)
13. Viganò, L.: Formal Methods for Socio-technical Security (Formal and Automated Analysis of Security Ceremonies). In: Coordination. LNCS, vol. 13271. Springer, Cham (2022)
14. Villers, A., et al.: Robot-assisted partial prostatectomy for anterior prostate cancer: a step-by-step guide. BJU Int. **119**(6), 968–974 (2017)
15. Weber, J., et al.: Effects of flow disruptions on mental workload and surgical performance in robotic-assisted surgery. World J. Surg. **42**, 3599–3607 (2018)
16. Weigl, M., et al.: Associations of intraoperative flow disruptions and operating room teamwork during robotic-assisted radical prostatectomy. Urology **114** (2018)
17. Zhou, X., et al.: Transvesical robot-assisted radical prostatectomy: initial experience and surgical outcomes. BJU Int. **126**(2), 300–308 (2020)

Operating Systems and Application Security

Flexible and Secure Process Confinement with eBPF

Carlo Mazzocca[ID], Andrea Garbugli[ID], Michele Armillotta[✉],
Rebecca Montanari[ID], and Paolo Bellavista[ID]

University of Bologna, Bologna, Italy
{carlo.mazzocca,andrea.garbugli,rebecca.montanari,
paolo.bellavista}@unibo.it, michele.armillotta2@studio.unibo.it

Abstract. To avoid potential bugs and vulnerabilities, it is crucial
to confine process execution within well-defined boundaries, specifying
which resources are accessible and what operations are allowed. Numerous technologies have emerged in Linux environments to address process
confinement or isolation. However, these solutions often lacked tailored
support, leading to a fragmented landscape of complex implementations.
Given the need to support different security abstractions, the Extended
Berkeley Packet Filter (eBPF) has emerged as a promising technology for
extending the capabilities of the Linux kernel functionalities, offering a
simple and flexible approach for process confinement. This paper introduces a framework that leverages eBPF to achieve flexible and secure
process confinement. We developed a prototype implementation and evaluated its overhead in limiting filesystem capabilities. Experimental findings underscore the effectiveness of our framework, demonstrating that
it can seamlessly integrate into Linux systems without incurring remarkable overhead.

Keywords: eBPF · Operating System Security · Application
Confinement · Process Isolation · Sandboxing · Linux Kernel Security

1 Introduction

Process confinement or isolation is a key security technique to prevent untrusted
components from directly accessing resources and executing operations. This is
particularly critical in contexts like cloud computing [18], where the integrity of
virtualized platforms and the software systems running on them can be compromised by malicious actors exploiting insufficient security isolation [2]. Thus,
confining processes within well-defined boundaries and restricting their access to
only authorized resources and operations reduces the potential damage that bugs
or vulnerabilities could cause [19], strengthening the overall system security.

Over the years, a wide variety of technologies for process confinement have
emerged on Linux (e.g., SELinux [4] and seccomp [15]). Despite their popularity,
most were not originally tailored for this task, and relying solely on individual

F. Martinelli and R. Rios (Eds.): STM 2024, LNCS 15235, pp. 97–109, 2025.
https://doi.org/10.1007/978-3-031-76371-7_7

mechanisms does not provide comprehensive security coverage. On the other hand, working solutions combine existing technologies, which can increase system complexity and introduce vulnerabilities [9]. Given the need to support different security abstractions without complex implementations, the Linux community introduced the Linux Security Modules (LSM) framework [20]. LSM offers a generic interface that allows the creation of flexible access control policies in a modular way, without modifying the kernel. Most of the confinement and isolation mechanics leverage kernel functionalities rather than developing kernel modules from stretch as it is a costly operation and small errors can result in serious failures.

In this direction, the Extended Berkeley Packet Filter (eBPF) [7] is envisioned as a valuable technology for implementing process confinement by running programs in the privileged context of the kernel without directly modifying it [6]. An eBPF program is loaded in the kernel as a bytecode and is activated when the kernel or an application triggers a certain hook point, e.g., the `write()` system call. Before being attached to the requested hook, an eBPF program is validated and the bytecode is translated into the machine-specific instruction set, achieving the same performance as natively compiled kernel code. Additionally, eBPF supports a data structure called *map* that allows efficient data exchange between userspace and kernelspace.

This paper presents a framework for implementing flexible and secure process confinement using eBPF. Our solution employs a modular architecture composed of various eBPF programs, each responsible for regulating access to different subsystems such as filesystem and networking. These programs enforce access control policies, provided through maps, which specify how processes are restricted. The combination of eBPF and access control policies allows dynamically managing Linux resources based on specified conditions.

We developed a prototype of the proposed framework and evaluated its performance in limiting filesystem capabilities, i.e., accessing protected resources. Experimental results demonstrate that eBPF can effectively confine processes attempting to access protected files in a Linux system, without introducing significant overhead. The primary contributions of this paper are as follows: (i) presenting a framework for implementing flexible and secure process confinement with eBPF, and (ii) developing a prototype and evaluating the overhead involved in restricting filesystem capabilities.

The remainder of this paper is structured as follows. Section 2 presents eBPF and offers an overview of existing security mechanisms in Linux systems. Section 3 reviews the related work in the field. In Sect. 4, we describe our framework, whose evaluation is reported in Sect. 5. Section 6 discusses experimental results and design choices. Finally, Sect. 7 concludes and presents future research directions.

2   Background

This section provides the background for understanding eBPF and the main mechanisms for implementing process confinement in Linux systems.

2.1   eBPF

As clearly outlined by the full form (i.e., Berkeley Packet Filter), the original BPF also referred to as classic BPF (cBPF) is a technology employed for packet filtering. However, the extended version can do much more and eBPF is considered a standalone term, often used interchangeably with BPF. The success of eBPF owes to its capacity to extend the kernel capabilities without modifying the kernel source code or loading kernel modules. This is achieved through eBPF programs, which run at the kernel level and add further capabilities to the operating system. These programs are event-driven as they are executed when the kernel or an application triggers a specific hook point, such as system calls or network events.

Thus, developers are only required to implement eBPF programs. To simplify this process, they can also write their eBPF programs using a subset of higher-level programming languages such as C by leveraging available development toolchains. For instance, our eBPF programs have been implemented in a C-like language and compiled through libbpf-bootstrap. These eBPF programs are associated with hooks and loaded into the kernel through bpf(2) system call. As eBPF programs are run in the kernel, they are restricted in what they can do and undergo a verification process that ensures they will not cause damage to the system. For instance, all programs must run to completion, hence, they cannot sit in a loop forever. To guarantee execution efficiency, the generic bytecode of the program is translated into the machine-specific instruction set through a Just-in-Time (JIT) compilation step. This has led to many novel use cases that leverage eBPF, including high-performance networking in cloud environments, and security functionalities as done in this paper.

2.2   Linux Process Confinement

Process confinement is a crucial concern for operating systems as it prevents unauthorized access to sensitive resources, ensures that processes operate within their designated boundaries, and provides better control and management of system resources (e.g., CPU, memory, and filesystem). In the following, we present the main mechanisms for process confinement in Linux systems and discuss their limitations.

Linux DAC: The most basic solution to confine processes is Discretionary Access Control (DAC) [20], which enforces security by ownership. A user owning a file can set the read, write, and execute permissions for that file. The system owner does not have full control over the system, while users control data at their discretion. In this access control model, the main concern is given by the root user who can do almost anything to a system, demanding additional and more secure mechanisms.

Linux MAC: When using Mandatory Access Control (MAC), the operating system constrains the capability of a process or user (including root) to access and perform operations on resources according to predefined security policies. Among existing LSMs implementing MAC, Security-Enhanced Linux (SELinux) and AppArmor are widespread solutions [4]. In SELinux, each process and resource is assigned a security context, used by security policies to grant or deny access. Policies define rules that establish how processes with security contexts can interact with resources having different security contexts. AppArmor allows administrators to specify the resources that are accessible from a process. Each process is bound to a profile that defines the set of permitted operations to resources. AppArmor profiles are simpler to write and manage, being more suitable for scenarios where simple MAC mechanisms are required.

Namespaces and Cgroups: Namespaces and groups are Linux features that limit the resources and how much of those resources a process can use [17], enabling a simple case of process confinement. Namespaces offer a mechanism to partition resources, allowing one group of processes to access a specific set of resources, while another group sees a distinct set. Partitioned resources span a spectrum: process IDs, file systems, network sockets, and user IDs. On the other hand, cgroups is another Linux feature for constraining non-enumerable resources like memory, CPU, and I/O bandwidth.

Ptrace: While primarily designed for debugging, the `ptrace` system call can be somehow employed for process confinement with certain limitations [5]. By leveraging this function, a process gains the ability to monitor and control its child processes, intercepting and manipulating system calls, modifying memory contents, or terminating the child process when required. However, employing `ptrace` for process confinement has significant drawbacks. These include notable performance overhead due to the need for system call interception, increased complexity in implementation, and increased security risks, as ptrace-based solutions may be vulnerable to privilege escalation or bypass.

Seccomp: Seccomp [15] is a security feature of Linux that restricts processes to a limited set of system calls, creating a secure and controlled execution environment. A process that tries to invoke a system call not included in the predefined set is terminated by the kernel. seccomp-bpf [13] is an extension that overcomes the low flexibility of seccomp by adding the capability to filter system calls through eBPF rules.

3   Related Work

Although eBPF has a history of focusing on performance, recent literature underscores a growing body of research that proposes it as a valuable technology to enhance system security. This section reviews key works utilizing eBPF for process confinement in Linux systems. Findlay et al. [9] proposed bpfbox, a tool that combines a policy language with eBPF to implement process confinement for userspace applications. While recognizing that this framework was among the first to adopt eBPF for process confinement, it does not support the instrumentation of processes at the system call level, a key feature of our framework. Additionally, from the publication date, bpfbox has not been maintained and new eBPF functionalities introduced since then are unsupported by this tool. Hung et al. [11] presented Sfiter, a framework for protecting security-critical kernel modules in Android devices. Sfiter uses predefined policies to filter system calls from untrusted processes to Qualcomm KGSL GPU device driver and Binder IPC modules. Although this tool can be potentially deployed on any device running Linux, it only focuses on these modules, resulting in limited flexibility compared to our solution. Jia et al. [12] created a programmable tool for filtering system calls called Seccomp-eBPF, which overcomes the restricted programmability of traditional seccomp where security policies are limited to static allow lists. However, seccomp-eBPF is restricted to system calls and does not consider hook points.

Moreover, using eBPF for process isolation has also found applications in specialized domains such as web applications. Abbadini et al. [1] introduced Cage4Deno, a Deno JavaScript and TypeScript runtime enhancement. This extension facilitates the creation of fine-grained sandboxes for executing subprocesses, safeguarding against compromised utilities that could threaten filesystem confidentiality and integrity. In containerization, eBPF has been adopted as a valuable technology to strengthen the host kernel against privilege escalation attacks mounted from within containers [8] and to allow containers to safely enforce fine-grained policies in the kernel [3].

4   Framework Architecture

As illustrated in Fig. 1, our framework's architecture consists of a policy manager, a loader, a policy enforcer, and several maps. The policy manager receives access control policies to enforce, parses them, and loads them into maps. Maps are particular eBPF data structures that connect the userspace applications and kernelspace eBPF programs. Once the loader compiles and loads the eBPF code into the kernel, the policy enforcer retrieves the access control policies from the eBPF maps. The policy manager and the loader are implemented in C, while eBPF programs are written in a restricted C-like language.

4.1   eBPF Maps

eBPF offers data structures called maps to facilitate information exchange between userspace and kernelspace [14]. These maps come in various types, such as hash, array, and bloom filters. Data shared using maps can be accessed

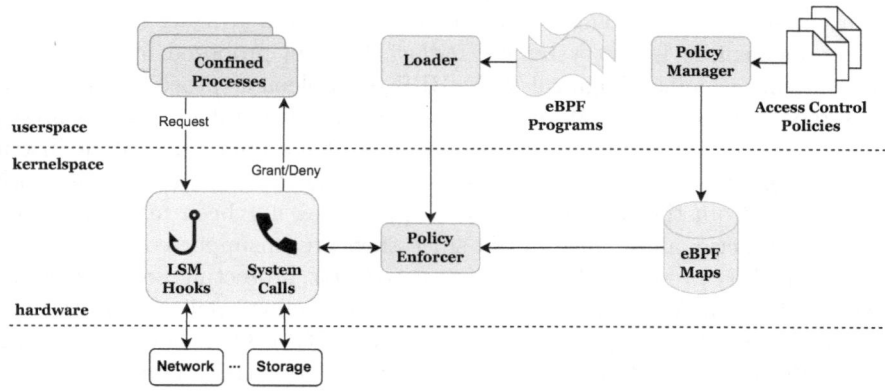

Fig. 1. Overview of our framework's architecture.

through the bpf(2) system call, which allows developers to create maps and efficiently manage their elements, i.e., read, update, and delete operations.

We use maps to share access control policies from userspace to eBPF programs running within the kernel. Maps are crucial to our architecture as they empower dynamic modification of access control policies without directly impacting eBPF programs. This flexibility allows for real-time adjustments to policy enforcement, enhancing the overall adaptability and security of the system.

4.2 Policy Manager and Loader

Access control policies are structured as key-value pairs, following the format key:value. To ensure high flexibility, each subsystem has policies that can only be defined and modified by authorized users (e.g., system administrators), as these policies are stored in a directory controlled by the root. We design policies following a *whitelisting* approach, where all operations and resources that are not specified are denied. By forcing authorized users to explicitly determine the admitted operations on resources, we prevent undesired circumstances where access to resources is unintentionally granted, e.g., creating a new file and forgetting to restrict its access.

The policy manager plays a key role in our architecture. It parses access control policies and stores them in maps, which are then referenced by the corresponding eBPF programs to enforce access control efficiently and flexibly. Listing 1.1 reports the snippet of an access control policy that restricts access to the file protected.txt exclusively to the cat program using the open() system call. Thus, any other operation will be rejected.

```
1  open.protectedFiles:/home/ex/Desktop/protected.txt;
2  open.allowedProc:cat;
```

Listing 1.1. Example of an access control policy.

It is worth noting that the proposed architecture decouples policy definition from their enforcement, allowing any changes without requiring modification the eBPF program source code.

Similarly to the policy manager that uploads policies into maps, the loader provides eBPF programs to the policy enforcer. The loader receives eBPF programs responsible for enforcing access control policies and compiles them into a loadable object file, which is then uploaded into the kernel.

4.3 Policy Enforcer

The policy enforcer consists of various eBPF programs, each responsible for regulating access to specific kernel functionalities and resources, such as the filesystem, networking, and process management. These eBPF programs are linked to a range of hooks within the Linux operating system, allowing them to intercept critical operations or events. Predefined hooks include system calls, kernel tracepoints, and network events, among others. When a process triggers a hook point, the corresponding eBPF program is activated and, based on the access control policies stored in maps, either grants or denies access to the resource or functionality.

Furthermore, our framework extends beyond simple authorization by implementing *controlled execution* of unauthorized processes. While existing solutions enforce access control policies by granting or denying access to certain resources, we allow unauthorized processes to perform mock executions, monitoring and logging their intended actions. This feature can be potentially useful for evaluating unknown process operations, effectively emulating a sandbox-like environment before letting them operate on the real system. For instance, regarding the write() system call, the policy enforcer can redirect it to a mock file, which can be analyzed later to assess potential threats.

It is worth noting that eBPF programs can be seamlessly attached to any predefined hook. If a predefined hook does not meet a particular need, developers can create custom kernel probes (kprobe) or user probes (uprobe) to intercept actions within kernel or user space applications. These capabilities empower users with great flexibility and fine-grained control over their operating systems.

In Listing 1.2, we report a snippet of an eBPF program that enforces the policy previously introduced. Specifically, this eBPF program restricts access to the file protected.txt according to the policy reported in Listing 1.1. Thus, only the cat program is allowed to execute the open() system call, while other operations are denied. For example, suppose a process attempts to perform a write() operation. In that case, the policy enforcer verifies whether the write() belongs to the permitted operations (lines 8–18) by comparing the system call invoked and the file path used by the process with the content of the corresponding map loaded by the policy manager. Since the write operation is not allowed on protected.txt, the value_proc is set to 0 and the access is denied. Additionally, to monitor intended operations, the write() system call is redirected to an error file (lines 24 and 25).

```
1  // Get file name
2  bpf_probe_read_user_str(filename, sizeof(filename), (char*)ctx->
       args[1]);
3
4  // Get process name
5  bpf_get_current_comm(comm, sizeof(comm));
6
7  // Look up elements in eBPF maps
8  value_file = bpf_map_lookup_elem(&OPEN_FILES_MAP, &filename);
9  value_proc = bpf_map_lookup_elem(&OPEN_PROC_MAP, &comm);
10
11 // Check flag status
12 if (value_file && (*value_file > 0)) {
13     flag_path = 1;
14 }
15
16 if (value_proc && (*value_proc > 0)) {
17     flag_process = 0;
18 }
19
20 if (flag_path == 1 && flag_process == 1) {
21     bpf_printk("Access denied\n");
22
23     // Redirect to error file
24     char new_path[MAX_DIM_PATH] = "/home/ex/Desktop/error.txt";
25     ret = bpf_probe_write_user((void *)ctx->args[1], new_path,
       sizeof(new_path));
26     return ret;
27 } else {
28     return 0;
29 }
```

Listing 1.2. Example of an eBPF program for process confinement.

5   Evaluation

In this section, we conduct experiments to evaluate how our framework impacts filesystem access latency. We developed a benchmark suite to measure the overhead imposed on both confined applications and the system when opening files. Our experimental findings demonstrate that our framework can enforce fine-grained access control with minimal overhead.

5.1   Experimental Setup

All experiments were performed on a local setup that matches a typical cloud environment. The machine runs Ubuntu 22.04 and is equipped with an 18-core

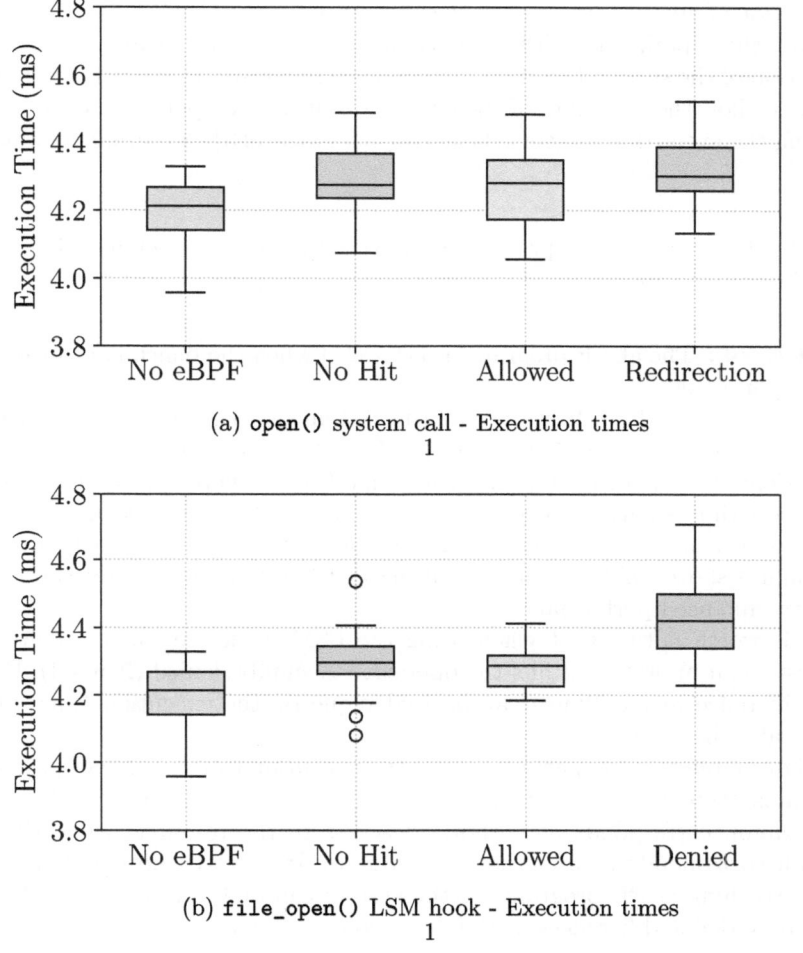

(a) `open()` system call - Execution times
1

(b) `file_open()` LSM hook - Execution times
1

Fig. 2. Performance evaluation of the proposed solution for different eBPF hooks (system call and LSM).

Intel i9-10980XE @ 3.00 GHz and 64 GB of RAM. We implemented a prototype of our framework in C and evaluated it using the eBPF program shown in Listing 1.2, which restricts access to a protected file.

5.2 Experiments

We conducted a set of experiments to evaluate the impact of our framework on the filesystem access latency. Each experiment was executed 50 times, and results have been averaged.

We evaluated the overhead imposed by our framework for authorized access to the file `protected.txt` (`Allowed`), access to an unprotected file (`No Hit`), and redirection of unauthorized access to the error file `error.txt` (`Redirection`). We also run the experiments without deploying our framework, using it as a baseline operation of the system (`No eBPF`) to compare the introduced overhead. Furthermore, to show the flexibility of the proposed solution, we performed experiments varying the event that triggers the activation of the eBPF program, summarized below.

System Call: The eBPF program is activated when the benchmark invokes the `open()` system call.

LSM Hook: The eBPF programs is activated when the benchmark invokes the `file_open` hook.

In Fig. 2a and Fig. 2b, we report the system's overhead associated with the `open()` system call and `file_open` LSM hook, respectively. Under both configurations, the operations do not show significant differences, with an average execution time slightly above 4 ms. Additionally, there are no notable differences among operations when using eBPF for enforcing access control policies. The figure shows that systems not using eBPF for confining processes do not achieve enhanced performance.

It is worth noting that when using the LSM hook, we cannot redirect the request to an unprotected file; the operation is simply denied (`Denied`). Despite this difference in handling unauthorized requests, the execution time remains practically the same.

Figure 3 offers a comparison among the execution time for kernelspace eBPF programs attached to the `open()` system call and `file_open` LSM hook. The `Hit` column corresponds to authorized access to the resource, while the `Miss` column corresponds to unauthorized requests. It is worth noting that, in this case, attaching eBPF programs at the system call or LSM hook level does not show remarkable differences as the execution time is a few microseconds.

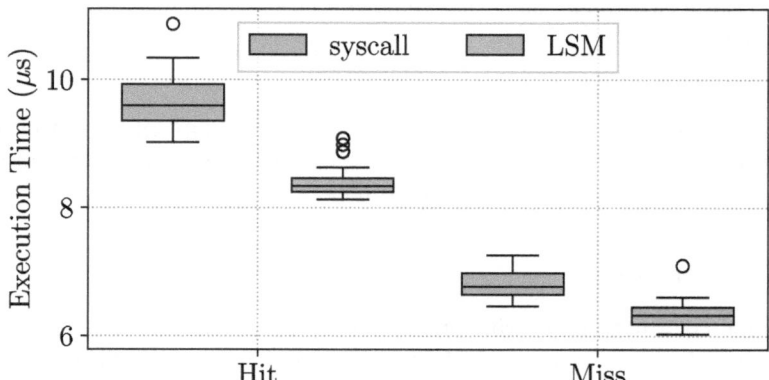

Fig. 3. Comparison of the execution times for the kernelspace eBPF programs attached to the `open()` system call and `file_open` LSM hook.

6   Discussion

Experimental results highlight that our eBPF-based framework can restrict filesystem access for opening files while introducing negligible overhead, regardless of whether the eBPF program is triggered by a system call or an LSM hook. The main difference observed between using the system calls and the LSM hooks levels lies in the flexibility they offer. Attaching the eBPF program to the open() system call enables controlled execution by redirecting the request to an error file. However, this method does not support returning an error code, which is possible when the eBPF program is triggered by the file_open hook. Moreover, system call redirection seems not always feasible, e.g., redirecting networking packets to a mock socket. In future work, we plan to deeply explore these mechanisms for the various Linux subsystems.

On the other hand, using the LSM hook allows the eBPF program to access more detailed information like the struct file, enabling the design of more fine-grained access control policies. For example, the file_open LSM hook can read the file's content and scan for forbidden keywords before allowing opening the file. However, this approach does not allow for redirecting the request to an error file as the primary role of this LSM hook is to allow or deny access, not to modify the behavior of the file operation to open a different file. This functionality may be achieved by creating a wrapper function that attempts to open the desired file and, upon receiving a denial, automatically redirects the request to an error file instead.

Given these considerations, it is evident that there cannot be a one-size-fits-all approach for selecting the level at which eBPF programs are triggered; the choice depends on the specific scenarios and requirements. For instance, LSM hooks are preferable when real-time information about the context of a potential threat is needed, and a simple yes-or-no decision suffices to determine whether the execution of the observed process is allowed. On the other hand, interception at the system call level is preferable when it is necessary to activate a redirect trigger to a mock and isolated "object" that can be analyzed later.

7   Conclusions and Future Work

This paper presents a framework for confining applications within Linux systems, leveraging the high flexibility of eBPF to enable fine-grained control over process behavior, enhancing security and system integrity. We implemented a prototype and evaluated its efficiency by measuring its impact on filesystem access latency. Experimental results demonstrate that using eBPF to confine processes introduces negligible overhead.

Future work includes expanding our framework to encompass additional critical Linux subsystems and comprehensively evaluating its impact across these areas. Furthermore, we plan to integrate our framework with network acceleration platforms to address security issues often overlooked in such contexts [10, 16]. Such frameworks typically leverage shared memory mechanisms to achieve zero-copy communications and enhanced performance, though this comes at the cost of removing isolation between applications. Integrating our eBPF solution for process confinement could address this issue.

Acknowledgments. This work was partially supported by the project SERICS (PE00000014) under the MUR National Recovery and Resilience Plan program funded by the European Union - NextGenerationEU.

References

1. Abbadini, M., Facchinetti, D., Oldani, G., Rossi, M., Paraboschi, S.: Cage4deno: a fine-grained sandbox for deno subprocesses. In: Proceedings of the 2023 ACM Asia Conference on Computer and Communications Security, ASIA CCS 2023, pp. 149–162. Association for Computing Machinery, New York (2023). https://doi.org/10.1145/3579856.3595799
2. Bazm, M.M., Lacoste, M., Südholt, M., Menaud, J.M.: Isolation in cloud computing infrastructures: new security challenges. Ann. Telecommun. **74**(3), 197–209 (2019)
3. Bélair, M., Laniepce, S., Menaud, J.M.: Snappy: programmable kernel-level policies for containers. In: Proceedings of the 36th Annual ACM Symposium on Applied Computing, SAC 221, pp. 1636–1645. Association for Computing Machinery, New York (2021). https://doi.org/10.1145/3412841.3442037
4. Brimhall, B., Garrard, J., De La Garza, C., Coffman, J.: A comparative analysis of linux mandatory access control policy enforcement mechanisms. In: Proceedings of the 16th European Workshop on System Security, EUROSEC 2023, pp. 1–7. Association for Computing Machinery, New York (2023). https://doi.org/10.1145/3578357.3589454
5. Connor, R.J., McDaniel, T., Smith, J.M., Schuchard, M.: PKU pitfalls: attacks on pku-based memory isolation systems. In: 29th USENIX Security Symposium (USENIX Security 2020), pp. 1409–1426. USENIX Association (2020). https://www.usenix.org/conference/usenixsecurity20/presentation/connor
6. Dejaeghere, J., Gbadamosi, B., Pulls, T., Rochet, F.: Comparing security in eBPF and WebAssembly. In: Proceedings of the 1st Workshop on EBPF and Kernel Extensions, eBPF 2023, pp. 35–41. Association for Computing Machinery, New York (2023). https://doi.org/10.1145/3609021.3609306
7. eBPF Documentation: eBPF (2024). https://ebpf.io/. Accessed 2 May 2024
8. Findlay, W., Barrera, D., Somayaji, A.: Bpfcontain: fixing the soft underbelly of container security. arXiv preprint arXiv:2102.06972 (2021)
9. Findlay, W., Somayaji, A., Barrera, D.: bpfbox: simple precise process confinement with eBPF. In: Proceedings of the 2020 ACM SIGSAC Conference on Cloud Computing Security Workshop, CCSW 2020, pp. 91–103. Association for Computing Machinery, New York (2020). https://doi.org/10.1145/3411495.3421358
10. Fried, J., et al.: Making Kernel bypass practical for the cloud with junction. In: 21st USENIX Symposium on Networked Systems Design and Implementation (NSDI 2024), pp. 55–73 (2024)
11. Hung, H.W., Liu, Y., Sani, A.A.: Sifter: protecting security-critical kernel modules in Android through attack surface reduction. In: Proceedings of the 28th Annual International Conference on Mobile Computing And Networking, MobiCom 2022, pp. 623–635. Association for Computing Machinery, New York (2022). https://doi.org/10.1145/3495243.3560548

12. Jia, J., et al.: Programmable system call security with ebpf. arXiv preprint arXiv:2302.10366 (2023)
13. Kernel, T.L.: Seccomp BPF (SECure COMPuting with filters) (2024). https://www.kernel.org/doc/html/v4.19/userspace-api/seccomp_filter.html. Accessed 2 May 2024
14. Miano, S., Bertrone, M., Risso, F., Tumolo, M., Bernal, M.V.: Creating complex network services with eBPF: experience and lessons learned. In: 2018 IEEE 19th International Conference on High Performance Switching and Routing (HPSR), pp. 1–8 (2018). https://doi.org/10.1109/HPSR.2018.8850758
15. manual page, L.: seccomp(2) (2024). https://man7.org/linux/man-pages/man2/seccomp.2.html. Accessed 2 May 2024
16. Rosa, L., Garbugli, A., Corradi, A., Bellavista, P.: INSANE: a unified middleware for QoS-aware network acceleration in edge cloud computing. In: Proceedings of the 24th International Middleware Conference, pp. 57–70 (2023)
17. Rosen, R.: Resource management: linux kernel namespaces and cgroups. Haifux **186**, 70 (2013)
18. Shu, R., et al.: A study of security isolation techniques. ACM Comput. Surv. **49**(3) (2016). https://doi.org/10.1145/2988545
19. Vahldiek-Oberwagner, A., Elnikety, E., Duarte, N.O., Sammler, M., Druschel, P., Garg, D.: ERIM: secure, efficient in-process isolation with protection keys (MPK). In: 28th USENIX Security Symposium (USENIX Security 2019), pp. 1221–1238. USENIX Association, Santa Clara (2019). https://www.usenix.org/conference/usenixsecurity19/presentation/vahldiek-oberwagner
20. Wright, C., Cowan, C., Smalley, S., Morris, J., Kroah-Hartman, G.: Linux security modules: general security support for the linux kernel. In: 11th USENIX Security Symposium (USENIX Security 2002) (2002)

Dynamic Analysis for Explainable Fine-Grained Android Malware Detection

Giovanni Ciaramella[1,2]($\boxtimes$) (iD), Francesco Mercaldo[2,3] (iD),
and Antonella Santone[3]($\boxtimes$) (iD)

[1] IMT School for Advanced Studies Lucca, Lucca, Italy
giovanni.ciaramella@imtlucca.it
[2] Institute for Informatics and Telematics, National Research Council of Italy
(CNR), Pisa, Italy
{giovanni.ciaramella,francesco.mercaldo}@iit.cnr.it
[3] University of Molise, Campobasso, Italy
{francesco.mercaldo,antonella.santone}@unimol.it

Abstract. Over the years, the frequency of cybersecurity attacks has surged as cybercriminals continually exploit vulnerabilities to amass profits through the unauthorized acquisition and resale of personal information on the dark web or by demanding ransoms. Fueled by this malicious motivation, researchers have diligently sought innovative methodologies to detect and thwart these cyber threats across various environments. Among the targeted landscapes, Android stands out due to its widespread usage, making it a prime target for attempted attacks. In response to this escalating challenge, in this paper we design and develop a method for identifying malicious and benign system calls through the usage of Deep Learning and an algorithm of Dynamic programming such as the Longest Common Subsequence algorithm. To conduct our research, we meticulously extracted System Calls from Android applications, transforming them into images to create a robust dataset comprising 13,570 samples. With the dataset in hand, we employed four different Convolutional Neural Networks, utilizing them to train and test various models. At the end of this process, our model achieved an accuracy rate of 0.890. To enhance the explainability of our findings, we applied two distinct Class Activation Mapping algorithms. These algorithms help spotlight the most significant areas during the classification process. Once these visual representations were obtained, we merged the original images with the heat maps generated by Class Activation Mapping algorithms. This fusion allowed to identify and extract the most discriminative system calls, providing valuable insights into the distinguishing features between malicious and benign behaviors.

Keywords: Security · Malware · Android · Explainability · Explainable Artificial Intelligence · Deep Learning · Artificial Intelligence

© The Author(s), under exclusive license to Springer Nature Switzerland AG 2025
F. Martinelli and R. Rios (Eds.): STM 2024, LNCS 15235, pp. 110–127, 2025.
https://doi.org/10.1007/978-3-031-76371-7_8

1   Introduction

Since the introduction of smartphones, tablets, computers, and personal devices into the market, these technologies have seamlessly become extensions of our daily lives. According to a study by StatCounter[1], as of early January 2024, approximately 60% (59.17%) of the global population owns a mobile phone, and this percentage is expected to continue rising. Unfortunately, this increased connectivity has provided cybercriminals an expanding opportunity to exploit. The rise in cybercrimes has been a persistent trend over the years, with malicious individuals increasingly targeting the theft of personal information for ransom or engaging in dark web sales. Android devices, in particular, have become prime targets due to their widespread use. According to a report published by StatCounter in December 2023[2], Android devices constituted a significant 70.48% of the global market share, overshadowing competitors like Apple, which held only 28.8%. Although Google introduced different methodologies to prevent the spread of malware, the latter always posed a threat even within the official store *i.e.,* Google Play[3]. As reported in an article published by Kaspersky in 2023[4] 620,000 applications downloaded were affected by trojan, and 1.5 million have been recognized as Chinese spyware in July 2023.

The original detection based on signature-based antimalware, relies on a database of known malware signatures to identify and block malicious software. While it has been a traditional and effective approach, it does have several weaknesses: as a matter of fact, it can only detect threats for which they have known signatures. If a new and previously unseen malware variant emerges, the signature-based approach may fail to recognize and block it. Moreover, this paradigm is ineffective against zero-day attacks, which exploit vulnerabilities that are not yet known to the security community [4]. Since there is no signature for these threats, they can go undetected until a new signature is created and distributed. The literature reflects the ongoing progress in research, introducing several additional methodologies for detecting malware or other dangerous applications. As reported in [9], to identify malware, researchers employed different methods like checking the permissions of an Android application, the opcode sequence, the API call, or studying the system call traces.

Starting from these considerations, in this paper, we propose a method to understand whether an application is malware or legitimate. We consider the proposed method as fine-grained because, it is not only able to mark an application under analysis as malware or trusted, but it can also automatically extract the system call sequence that, from the model point of view, is symptomatic of the malware/trusted behavior.

In a nutshell, the proposed method for malware detection relies on dynamic analysis *i.e.,* we extract the system call sequence from running applications.

---

[1] https://gs.statcounter.com/platform-market-share/desktop-mobile-tablet/worldwi de/yearly-2023-2024.

[2] https://gs.statcounter.com/os-market-share/mobile/worldwide.

[3] https://play.google.com/.

[4] https://www.kaspersky.com/blog/malware-in-google-play-2023/49579/.

Thus, for each application an image is built directly from the system call sequence: these images (one image for each application) represent the input for a set of convolutional neural networks [5–7,11]. Once the models are built, we consider the best one to extract, through two different Class Activation Mapping (CAM) algorithms, the area that from the model of view is most discriminative for the malware prediction. Considering that the images are built starting from the system call sequences, the areas highlighted from the CAMs can be traced back to system calls, for this reason, we apply a Longest Common Subsequence (LCS) algorithm to find the maximum common system call sequences to malicious applications. For this reason, we consider the proposed method fine-grained, considering its ability to identify the malicious and benign sequence in terms of system calls. We consider dynamic analysis, with particular regard when system calls are considered as features, considering that dynamic analysis is less prone to be eluded by obfuscation techniques, as demonstrated in [10].

The paper proceeds as follows: the next section shows an overview related to the state-of-the-art related to adoption of dynamic analysis for Android malware detection, in Sect. 3 the proposed method is presented, the results of the experimental evaluation, performed on real-world Android applications, is shown in Sect. 4 and, finally, conclusion and future research lines are drawn in the last section.

2   Related Work

Over the years, cybersecurity has become one of the most significant issues in several fields like Cloud Security, the Internet of Things (IoT), and Mobile Security. In the latter, cybersecurity seeks to safeguard data stored on devices, including tablets and smartphones. Different researchers started to study several methods to address potential software bugs that could be exploited by malicious users, leading to data leaks. In Table 1, we have presented a structured literature review, intended to enable comparative analysis with papers encompassing various studies, methodologies, and outcomes.

Dimjašević et al. in [2], proposes a methodology named MALINE to recognize malware and trusted applications employing Machine Learning. In their method, they extracted system calls on 12,000 Android applications and, after this step, they applied different ML algorithms using R as a programming language. The experimental outcomes revealed an initial accuracy of 93%, increased to 96% up-sampling techniques. In contrast to the authors, we abstain from employing sandboxes. A sandbox serves as a secure and isolated environment for monitoring an application's actions without impacting the actual system. Our decision to forgo their use is rooted in the intention to directly observe unusual behaviors manifested by the operating system following specific actions.

Authors in [14] adopted a two-step feature selection approach based on Rough Set and Statistical Test (RSST) to extract refined system calls, able to categorize malware and benign applications with a higher accuracy (99.9%). Kumar et al. employed static and dynamic features like permissions, opcodes, APIs, system

Table 1. Comparative analysis with previous works.

Previous Works	Year	Architecture	XAI	Syscall Iden.	Accuracy
Dimjašević et al. in [2]	2016	Android	✗	✗	0.960 SUD
Kumar et al. in [14]	2019	Android	✗	✓	0.999 HD
Xiao et al. in [17]	2019	Android	✗	✗	0.966 BD
Surendran et al. in [13]	2020	Android	✗	✓	> 0.900 BD, SUD ~0.720 HD
Mercaldo et al. in [10]	2023	Android	✓	✗	0.720 BD 0.730 BD
Gupta et al. in [3]	2024	Android	✗	✗	0.975 SUD
Our Method	**2024**	**Android**	✓	✓	**0.890 BD**

LEGEND:
XAI: Explainability, **Syscall Iden.:** System Call Identification, **BD:** Balanced Dataset,
SUD: Slightly Unbalanced Dataset, **UD:** Unbalanced Dataset, **ND:** Not Described
HD: Highly Unbalanced Datasets

calls, network traces, and system call graphs to conduct the analysis. Different from them, we focused our research paper on identifying malicious/benign System Calls to understand which combination of them can be used by malicious users.

In [17] researchers introduced an approach to malware detection on Android platforms by leveraging Deep Learning techniques. In detail, they employed the Long Short-Term Memory (LSTM) language model to discern malicious applications from trusted ones. The classifier proposed reached an accuracy of 96.6% and utilizes these scores to ascertain the nature of the application under examination, determining whether it is malicious or trustworthy. The decision is based on the superiority of the two computed scores. Different from Xiao et al., we employed several Deep Learning architectures retrieved from the literature, and to train and test them, we converted the system calls obtained from each application into images using a Python script written by the authors.

Surendran et al. in [13] introduced a tool designed to identify malicious Android applications based on System Calls. The authors augmented their approach by integrating a Machine Learning classifier. Impressively, the average accuracy achieved reached 0.955, showcasing the effectiveness of their methodology. Even in the case of a highly unbalanced dataset, the obtained precision remained commendable at 0.720, underlining the robustness of their system. Distinct from them, we employed a methodology based on Deep Learning with an accuracy value of 0.890 using a dataset composed of augmented images in the training set.

Authors in [10] presented an approach to demonstrate the efficacy of dynamic analysis and Deep Learning in constructing robust models. In detail, researchers proposed two experiments using a convolutional neural network built by them. In the first experiment, the model was trained and tested using a dataset comprised of known malware samples. In the second experiment, the model is trained on the malware dataset but evaluated using a dataset composed of obfuscated malware. In conclusion, the analysis proposed of the malware and obfuscated detection models, examining their resilience through the lens of explainability.

Gupta *et al.* [3] developed a method to identify malware in the Android environment employing the usage of permissions and system call pairs. In detail, in their study, the authors combined the advantages of static and dynamic analysis obtaining an accuracy score of 0.975. Different from them, we proposed a method able to classify an image in malware or trusted and after that retrieve the most important system calls that allowed that classification. To do that, we employed two different CAM algorithms.

3   The Method

In this section, we provide an overview of the proposed method to identify the most common system calls employed by malware and trusted applications. In this proposed method, we extract the system call traces from several running applications, and from them, we obtain images as a result. After this phase, we trained and tested five Convolutional Neural Networks architectures (available in the literature) to classify the images obtained. As the last steps, we also employed two different CAM algorithms to identify the areas of the image responsible for a specific prediction. From the heatmap obtained, we retrieved the Longest Common Subsequence of System Calls employed by both types of applications *i.e.,* Malware and Trusted. In Fig. 1 all steps of the proposed method are reported.

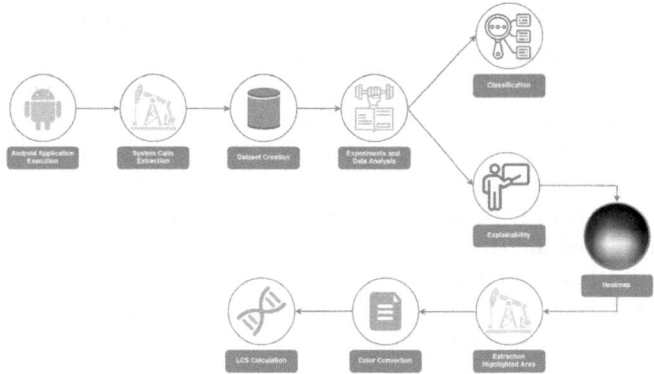

Fig. 1. The proposed method for obtaining the most common system calls used by malware and trusted applications.

3.1   System Call Extraction

The first step of the proposed methodology was the System Call Extraction. To perform that, we captured and stored (in textual format) the system calls generated by running applications. Once obtained a set of Malware and Trusted Android applications in *.apk* format, we executed them on an Android device emulator, generating a different event every 10 s. At the end of this process, we sent 25 operating system events on each application installed and executed on the device. The decision to employ this set of system events comes from detailed research published in the literature, where the authors demonstrate their maliciousness. In detail, these events enable malicious payloads to be activated in the Android environment. The operations mentioned above have been automatically performed employing a script written by the authors intended to carry out a set of actions like nitialize the Android emulator, install the .apk, capture system call traces, sequentially send and process 25 OS events, save the trace, terminate the application and emulator, and revert the emulator disk to a clean snapshot.

After the possible malign system call trace acquisition, we also employed a tool to generate a pseudo-random user event like gestures or touches. During the usage of the tool named *Android Debug Bridge*[5] (ADB) (*v. 1.0.32*), we used *STRACE*[6] to collect the system call. The latter is freely available on Linux operating systems, and invoking the command `strace -s` PID made it possible to intercept only the system calls generated by the application under the analysis process during the execution of the Android application under analysis. Obtained the possible malicious and benign system calls from each application tested, we needed to convert them into images. The latter method has been selected for training Convolutional Neural Networks, *i.e.,* a type of Deep Learning (DL) model explicitly crafted for processing and analyzing visual data, such as images. The approach involved constructing the set of images through pixel-by-pixel methods to create our model. In detail, we created a bi-dimensional vector (*i.e.,* matrix) of various dimensions (depending on the number of system calls retrieved), and each pixel was assigned the RGB value previously associated with each system call. As a result of the process mentioned above, we obtained various images, with two examples depicted in Fig. 2. In detail, Fig. 2a reports an example of images obtained from the system call traces. Furthermore, for confirmation of the file's malicious origin (Hash: `fe5fe15cde9f9271c31e686582307a2f3bad8f09aafdee942a5d3497107a4d22`), we verified its hash code on VirusTotal[7] - a free online service specifically designed for the analysis of files and URLs to identify and detect malicious content - where 34 antimalware on 65 categorized it as malware. Instead, Fig. 2b shows an example of a trusted application. The latter belongs to the application "com.eurosport" and is available on Google Play.

3.2 Dataset

Creating an acceptable dataset is one of the most significant components in obtaining effective results. In this paper, we built the final dataset by employing two different families: malware and trusted. The first was populated using

---

[5] https://developer.android.com/tools/adb.
[6] https://man7.org/linux/man-pages/man1/strace.1.html.
[7] https://www.virustotal.com.

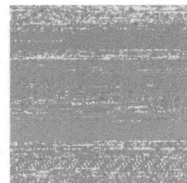 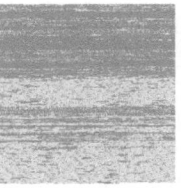

(a) Example of image retrieved from malware *apk* (HASH: fe5fe15cde9f9271c31e686582307a2f3bad8f0 9aafdee942a5d3497107a4d22).

(b) Example of image retrieved from trusted *apk* (Android Package: com.eurosport).

Fig. 2. Example of two different types of images belonging to Malware and Trusted applications.

the *Derbin Dataset* [1], while legitimated applications were retrieved from the official app store of Google using an open-source crawler. The *Derbin Dataset*[8] is widely used in the malware detection field thanks to its composition and the variety of families it consists of. In the pursuit of comprehensively addressing various mobile malicious behaviors, the experimental evaluations encompassed the examination of the following categories of malware families: *Geinimi, Plankton, Basebridge, Kmin, GinMaster, Opfake, FakeInstaller, DroidDream, DroidKungFu,* and *Adrd.* Those families represent the ten most populous families of the dataset. On the other hand, to build the trusted dataset, we downloaded them from Google Play employing a crawler. Moreover, to verify the trustworthiness of the applications downloaded from Google Play[9], we also submitted those apps to Virustotal[10].

3.3   Experiments

In this section, we provide an overview of the Convolutional Neural Networks employed in this research article, also explaining the hyperparameters used. Moreover, in Sect. 3.3 the different architectures utilized are shown, while in Sect. 3.3 we furnish more details about the explainability algorithms adopted.

Classification. Upon acquiring the dataset, we undertook a comprehensive exploration, conducting a series of experiments using five state-of-the-art Convolutional Neural Networks (CNNs): *MobileNet, Resnet 50, Standard CNN,* and *VGG19.* These experiments sought to assess the performance of each architecture in the context of distinguishing between malware and trusted instances. To ensure a thorough evaluation, we employed a variety of hyperparameters during the training and testing phases. The hyperparameters, as outlined in Table 2, were extracted from the results of the most successful experiment.

---

[8] https://drebin.mlsec.org/.
[9] https://play.google.com/.
[10] https://www.virustotal.com/gui/.

Table 2. Hyperparamters adopted to train models

Model	Image Size	Batch Size	Epochs	Training Time HH:MM:SS
MobileNet	110×3	32	40	00:08:26
ResNet50	110×3	64	20	00:07:24
	110×3	64	25	00:09:00
Standard CNN	110×3	64	30	00:02:18
VGG19	110×3	64	35	00:16:16

Specifically, we maintained a consistent image size of 110×3 pixels across all experiments, facilitating a more accurate and insightful comparison of outcomes. The training process involved the exploration of various epoch values, including 20, 25, 30, 35, and 40, coupled with two distinct batch sizes: 32 and 64. This meticulous approach allowed us to scrutinize the impact of different hyperparameter configurations on the performance of the selected CNN architectures, providing valuable insights into their efficacy in classifying instances as either malware or trusted.

Explainability. Following the evaluation of the models, the subsequent stage denoted as the *Explainability* in Fig. 1, involves the generation of two different CAMs through the utilization of the Gradient-weighted Class Activation Mapping (Grad-CAM) [12], and Score-weighted Class Activation Mapping (Score-CAM) [15] algorithms. Those algorithms serve a dual purpose: they provide a visual explanation for the model's predictions and aid in debugging. The intention behind this step is not only to achieve high prediction accuracy but also to scrutinize specific areas of the image (i.e., the application) that contribute to a particular prediction. The objective is to shed light on the prediction process by identifying the specific regions within the images that contribute to it. Following this identification, the focus shifts to retrieving system calls, allowing for an understanding of the LCS between malware and trusted applications. This functionality provides security analysts with the means to assess whether the model effectively concentrates on pertinent regions to make accurate predictions and also to allow them to determine which sequence of system calls can be blocked to avoid malicious activities.

3.4 Longest Common Subsequence

As the last step of the proposed methodology, we exploit the Longest Common Subsequence (LCS) to identify that in the common subsequence of malware and trusted. This type of algorithm belongs to Dynamic Programming (DP), which solves different problems by dividing the problems into subproblems. Unlike other algorithmic approaches, DP preserves the solutions to individual subproblems in a table avoiding the re-calculation whenever the subproblem recurs.

About the LCS, the latter is commonly used in biology to compare two or more organisms to determine how closely related these are. As defined in the Optimal substructure for LCS's theorem [8]: If $X = \langle x_1, \ldots, x_m \rangle$ and if $Y = \langle y_1, \ldots, y_n \rangle$ are sequences, let $Z = \langle z_1, \ldots, z_k \rangle$ be some LCS of X and Y.

- If $x_m = y_n$, then $z_k = x_m$ and Z_{k-1} is an LCS of X_{m-1} and Y_{n-1}.
- If $x_m \neq y_n$, then $z_k \neq x_m$ - Z is an LCS of X_{m-1} and Y.
- If $x_m \neq y_n$, then $z_k \neq y_n$ - Z is an LCS of X and Y_{n-1}.

To identify the LCS of the System Calls contained in the malware and trusted samples analyzed, we employed several Python scripts executed for each image contained in the test folder. As a first step, we retrieved the highlighted System Call from the CAMs algorithm, we identified the position of the pixel, and thanks to a legend created before we retrieved the system call associated with the color. As the last step, we calculated the Longest Common Subsequence. To perform the latter stage an algorithm is executed in polynomial time $i.e.$, $O(m \cdot n)$ where m and n are the lengths of the input strings.

4   Experimental Analysis

In this section, we discuss the results obtained after the experimental analysis. More in detail, Sect. 4.1 shows the outcomes of the models obtained; in Sect. 4.2 we discuss the results achieved using CAMs algorithm., while in Sect. 4.3 we describe the Longest Common Subsequence obtained after the analysis of *malware*, and *trusted* application. The following hardware and software characteristics were employed for the execution of the studies in this study paper: CPU Intel Core i7 12th Generation, GPU NVIDIA GeForce RTX 3060 Ti (8 GB), RAM DDR4-SDRAM 3200 MHz (16 GB), and the operating system used was the Ubuntu 22.04 LTS distro.

4.1   Models Evaluation

Before the execution of the experiment, we needed to create a dataset. As specified in Sect. 3.2 we employed the *Derbin Dataset* to obtain several malware applications in *.apk* format, while to obtain the trusted applications, we downloaded them from Google Play employing a crawler. In detail, we used **3,360** malware application and **3,468** trusted. Due to the poor accuracy obtained with this dataset (less than 82%), we decided to employ the *augmentation technique* to train several Convolutional Neural Networks using a dataset composed of an increased number of samples. Typically, various transformations, such as *rotation, scaling, translation, shearing, zooming, brightness and contrast adjustment*, and others, can be implemented during this process. Different from them, in our approach, we exclusively applied horizontal and vertical flipping for each image. At the end of this process, we obtained a dataset composed of **7,414** malware and **7,520** trusted for a total of **14,934** images. After the augmentation process, the dataset was partitioned into three subsets with an 80-10-10 distribution, comprising 13,570 samples in the training set, 682 samples in the validation set, and 682 samples in the test set. Of course, the samples obtained through the augmentation technique were employed only in the training and not during the validation and testing phases.

Table 3. Results obtained after test set execution

Model	Loss	Accuracy	Precision	Recall	F-Measure	AUC
MobileNet	0.671	0.878	0.878	0.878	0.878	0.924
ResNet50	0.429	0.891	0.891	0.891	0.891	0.941
	0.457	0.890	0.890	0.890	0.890	0.948
Standard CNN	0.525	0.886	0.886	0.886	0.886	0.930
VGG19	0.693	0.507	0.507	0.507	0.507	0.507

Once trained CNNs with the hyperparameters reported in Table 2, we proceed to the testing obtaining the results shown in Table 3. As can be denoted in the latter Table, the ResNet50 architectures reached higher outcomes in terms of accuracy and precision. In detail, we reported in Table 3 two different results, the first one trained with a batch size of 64, several epochs of 20, and a learning rate of 0.0001, and the second one with a batch size of 64, 25 epochs and learning rate of 0.0001. To check the effectiveness of the classification, we applied a specific study of the training and the validation phases, trying to identify the minor loss and the best accuracy during these stages. This study allowed us to identify the second experiment as the best one.

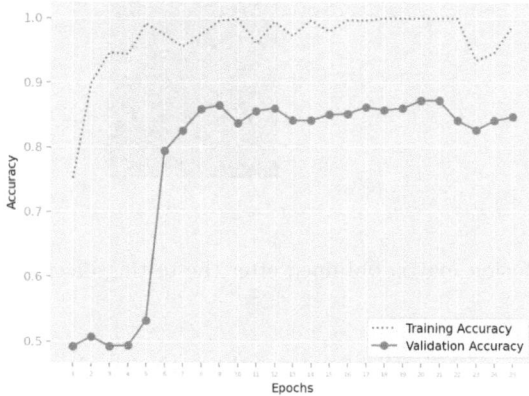

Fig. 3. Accuracy achieved in each epoch during the training and validation phases using the ResNet50 architecture.

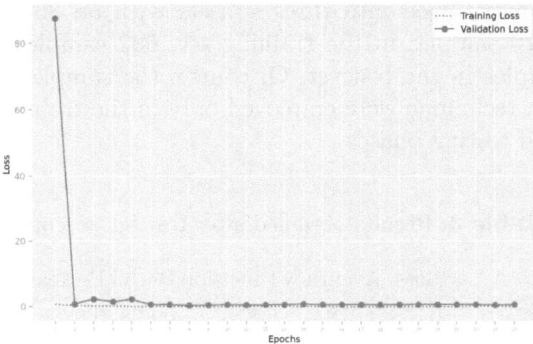

Fig. 4. Loss achieved in each epoch during the training and validation phases using the ResNet50 architecture.

To explain the outcomes retrieved, Figs. 3, and 4 report the trend of the accuracy, and the loss during the phases that belong to the best model obtained. In detail, in these figures, the accuracy/loss values obtained during the training are reported using a red line, while the accuracy/loss reached at each epoch during the validation in blue. After the training and test phases, we also calculated the confusion matrix for the best model obtained.

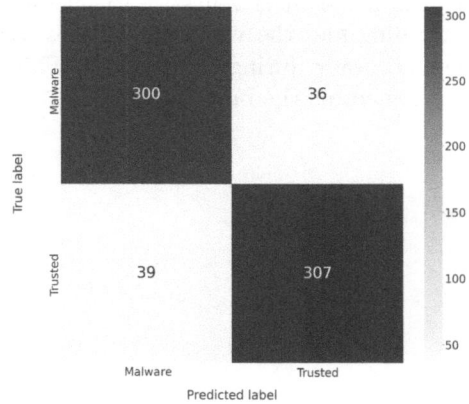

Fig. 5. Confusion matrix obtained after the testing phase of ResNet50.

The latter, reported in Fig. 5, shows how the model can identify trusted and malware samples in most cases. In the evaluation of a binary classification model distinguishing between malware and trusted instances, the confusion matrix reveals insightful performance metrics. For the malware class, the model achieved 300 True Positives (TP), correctly identifying malicious instances, and

307 True Negatives (TN), accurately classifying trusted instances. However, there were 39 False Positives (FP), instances wrongly classified as malware, and 36 False Negatives (FN), where the model failed to recognize actual malware. The corresponding metrics reflect a balanced performance: an accuracy of 0.890, indicating the overall correctness of predictions; a precision of 0.885, highlighting the reliability of positive predictions; a recall of 0.892, emphasizing the model's ability to capture actual malware instances; an F1 score of 0.888, representing the harmonic mean of precision and recall; and an area under the curve (AUC) of 0.890, quantifying the model's ability to discriminate between malware and trusted instances. The metrics collectively depict a well-performing model, effectively balancing the trade-off between precision and recall in classifying instances as either malware or trusted. Other models trained and tested reached lower results, in particular VGG19. As reported in Table 3, the latter demonstrated suboptimal performance, failing to exceed the threshold of 0.507 in any of the metrics assessed.

4.2 CAMs Application

Once trained and tested different architectures and retrieved the best model (ResNet50 according to Sect. 4.1), we applied two different CAM algorithms to the latter. In detail, we decided to use Grad-CAM and Score-CAM algorithms to understand if the same areas are highlighted - same System Calls in this case - despite the different methodologies employed. Although both methods aim to visualize the important regions that contribute to a model's decision, the difference is in the computation of the gradients. Specifically, in Grad-CAM, gradients are calculated concerning the feature maps of the final convolutional layer. The importance weights are determined using global average pooling. On the other hand, Score-CAM computes gradients concerning the input image, directly utilizing them to weight the raw class scores. Additionally, Score-CAM employs a Rectified Linear Unit (ReLU) activation to emphasize positive contributions. Following the application of CAMs, we acquired a set of images in PNG format where the activation map is overlaid onto the original image. The heatmaps are designed to account for three distinct shades: yellow, green, and blue. Specifically, regions shaded in blue indicate areas deemed less relevant to the model. Conversely, the model expresses interest in the yellow-colored portions of the image. Green areas, on the other hand, serve to highlight central regions. Figures 6 and 7 report the application of Grad-CAM and Score-CAM on a malware sample with hash 1ffcae9872d1707c058d7a50ffb44bc1809af08d5c57949510782827b082 f1f5 which belongs to the *Trojan* family. Both algorithms highlight the same areas of the image (in the center), and also the classification reached a high accuracy level of 100%.

Figures 8 and 9 report the application of both CAM algorithms on trusted samples. Again, the accuracy turns out to be 100%, and the areas highlighted are the same.

In summary, the application of Grad-CAM and Score-CAM algorithms proves to be intriguing for classifying and differentiating a set of images. Notably, the

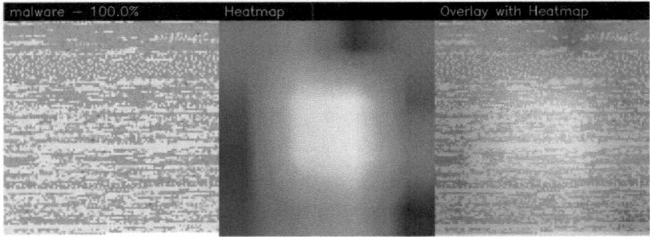

Fig. 6. Result of applying Grad-CAM algorithm on malware sample (Hash: 1ffcae9872d1707c058d7a50ffb44bc1809af08d5c57949510782827b082f1f5).

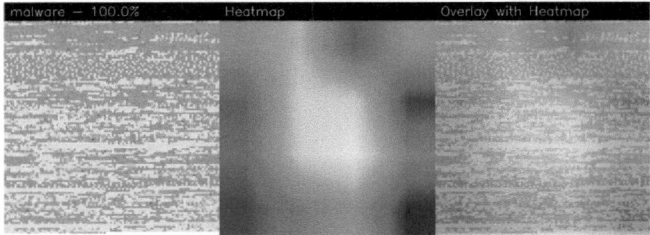

Fig. 7. Result of applying Score-CAM algorithm on malware sample (Hash: 1ffcae9872d1707c058d7a50ffb44bc1809af08d5c57949510782827b082f1f5).

incorporation of explainability allows users to identify the most distinctive region within an image. In addition, the usage of those algorithms allowed us to proceed to the last step of this research paper: the identification of the Longest Common Subsequence of malware and trusted samples.

4.3   Longest Common Subsequence

The final step of the proposed methodology was obtaining the Longest Common Subsequence of System Calls from malware and trusted samples taken into account during the test phase. Applied the CAMs algorithms, we retrieved the heatmap of the most interested areas by the model. From that, we used a Python script written by the authors to recover the RGB colors previously associated with System Calls. Got the name of the system call, we executed an LCS algorithm. We performed the previously outlined steps for both CAM algorithms, aiming to comprehend the similarities in the detection process. This concluding phase aims to help the community identify malware and trusted applications using a specific pattern. The usage of LCS can also contribute to forging a signature for the applications to improve the detection mechanism. In addition, using a specific Longest Common Subsequence application can identify a precise family or category. In this Section, we provide the results obtained after the algorithm's execution for both classes employed in this research article (*i.e.,* malware and trusted), and in addition, we also try to furnish an explanation of

Fig. 8. Result of applying Grad-CAM algorithm on trusted sample (ch.srf.radioapp).

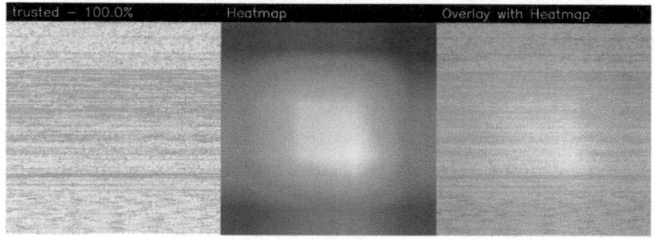

Fig. 9. Result of applying Score-CAM algorithm on trusted sample (ch.srf.radioapp).

why the use of specific system calls analyzing each system call identified using man[11] webpage.

The model effectively identifies malware samples, emphasizing distinct regions within the sample's image use that distinguish these according to the family they belong to. In Listing 1 are reported the System Calls obtained from the areas identified by the Grad-CAM algorithm, while the System Calls obtained using the Score-CAM are shown in Listing 2. Both retrieved LCSs is possible are identical. That means that each CAM algorithm has identified the same areas. The System Calls identified can be resumed in *futex, clockgettime, mprotect, sigprocmask,* and *cacheflush.*

Listing 1. Longest Common Subsequence retrieved on malware samples using Gradient-weighted Class Activation Mapping algorithm.

```
futex x8 -> clock_gettime x1 -> futex x1 -> mprotect x1 ->
cacheflush x1 -> mprotect x1 -> futex x5 -> sigprocmask x1 ->
mprotect x2 -> cacheflush x1 -> futex x6 -> sigprocmask x1 ->
mprotect x1 -> cacheflush x1 -> mprotect x1 -> futex x8 ->
mprotect x2 -> cacheflush x1 -> futex x6 -> mprotect x2 ->
cacheflush x1 -> mprotect x1 -> futex x7 -> mprotect x2 ->
cacheflush x1 -> futex x1 -> mprotect x1 -> cacheflush x1 ->
futex x7 -> mprotect x2 -> cacheflush x1 -> mprotect x1 ->
futex x6 -> cacheflush x1 -> mprotect x1 -> futex x2
```

---

[11] https://man7.org/.

Listing 2. Longest Common Subsequence retrieved on malware samples using Score-weighted Class Activation Mapping algorithm.

```
futex x8 -> clock_gettime x1 -> futex x1 -> mprotect x1 ->
cacheflush x1 -> mprotect x1 -> futex x5 -> sigprocmask x1 ->
mprotect x2 -> cacheflush x1 -> futex x6 -> sigprocmask x1 ->
mprotect x1 -> cacheflush x1 -> mprotect x1 -> futex x8 ->
mprotect x2 -> cacheflush x1 -> futex x6 -> mprotect x2 ->
cacheflush x1 -> mprotect x1 -> futex x7 -> mprotect x2 ->
cacheflush x1 -> futex x1 -> mprotect x1 -> cacheflush x1 ->
futex x7 -> mprotect x2 -> cacheflush x1 -> mprotect x1 ->
futex x6 -> cacheflush x1 -> mprotect x1 -> futex x2
```

Similar to malicious software applications, the system calls extracted from trusted samples are shown in Listings 3 and 4. In detail, in Listing 3 are reported system call retrieved after the Grad-CAM algorithm application, while in Listing 4 the System calls obtained after Score-CAM algorithm execution. As can be denoted from them, both CAM algorithms identified the same area and the same system calls (*futex* and *mprotect*) as a consequence. The only difference is the presence of the `futex` system call (Grad-CAM 20 appearances and Score-CAM 21).

Listing 3. Longest Common Subsequence retrieved on trusted samples using Gradient-weighted Class Activation Mapping algorithm.

```
futex x20 -> mprotect x6
```

Listing 4. Longest Common Subsequence retrieved on trusted samples using Score-weighted Class Activation Mapping algorithm.

```
futex x21 -> mprotect x6
```

Over the years, different papers have been published in this field, also considering System Calls to discriminate benign and malign applications over different Operative Systems. One of the first studies published on that was proposed by Xiao *et al.* in 2016 [16]. In that article, researchers display a list of the most frequent system calls employed in malware and trusted applications in the Android environment. Despite past years, the system calls used to turn out to be the same in many cases; what changes, however, turn out to be the techniques used. In this article, after the LCS algorithm execution, we identified 5 System Calls that can be used for benign or malicious purposes depending on the use. According to the man webpage, `futex` is typically used as a blocking construct to synchronize shared memory. Developers employ primitives like mutexes (*i.e., mutual exclusion*) or condition variables to coordinate access to shared resources. About `clockgettime`, this type of system call is employed to retrieve the current time with a high-resolution timestamp to obtain precise time information. While `mprotect` has the aim to change the protection of a memory region, thus allowing to mark an area of memory as read-only, read-write, or execute-only. The `sigprocmask` system call is utilized to control the signal mask, determining which signals are currently blocked (ignored) and which are allowed for a given process. This functionality enables applications to manage signal handling during critical sections of code, ensuring that interruptions by signals are controlled and predictable. Typically used in conjunction with signal handling mechanisms, sigprocmask enhances a program's ability to maintain stability and responsiveness in the face of asynchronous signals. The final system call identified is `cacheflush`, designed to clear the instruction and/or data cache for a specified range of memory addresses. This system call finds its typical application in situations where maintaining cache coherency is essential, such as when dynamically modifying executable code during runtime or working with memory-mapped I/O regions. The System Calls identified and highlighted by the CAMs algorithms into the malware samples (*futex, clockgettime, mprotect, cacheflush*, and *sigprocmask*) could potentially be used for malicious activities, particularly related to security evasion, privilege escalation, or exploiting vulnerabilities.

5   Conclusion and Future Work

In this paper, we introduced a novel malware detection approach centered around the analysis of System Calls. Specifically, we curated a dataset comprising images captured during the execution of system calls and pseudo-random user events. Following the dataset creation, we leveraged various deep-learning models for classification. Upon identifying the optimal model, we applied two distinct CAM algorithms (*i.e.,* Grad-CAM and Score-CAM) to identify System Calls utilized in malware and trusted samples. As a result, we obtained a list of 5 system calls typically exploited for malicious actions. In future work, we would to increase the dataset and avoid using augmentation. In addition, we intend to extend the work to detecting the actions done by the system call to understand the changes performed by them.

Acknowledgment. This work has been partially supported by EU DUCA, EU Cyber-SecPro, SYNAPSE, PTR 22–24 P2.01 (Cybersecurity) and SERICS (PE00000014) under the MUR National Recovery and Resilience Plan funded by the EU - NextGenerationEU projects, by MUR - REASONING: foRmal mEthods for computAtional analySis for diagnOsis and progNosis in imagING - PRIN, e-DAI (Digital ecosystem for integrated analysis of heterogeneous health data related to high-impact diseases: innovative model of care and research), Health Operational Plan, FSC 2014–2020, PRIN-MUR-Ministry of Health, the National Plan for NRRP Complementary Investments D^3 4 Health: Digital Driven Diagnostics, prognostics and therapeutics for sustainable Health care, Progetto MolisCTE, Ministero delle Imprese e del Made in Italy, Italy, CUP: D33B22000060001 and FORESEEN: FORmal mEthodS for attack dEtEction in autonomous driviNg systems CUP N.P2022WYAEW.

References

1. Arp, D., Spreitzenbarth, M., Hubner, M., Gascon, H., Rieck, H., Siemens, C.E.R.T.: Drebin: effective and explainable detection of android malware in your pocket. In: NDSS, vol. 14, pp. 23–26 (2014)
2. Dimjašević, M., Atzeni, S., Ugrina, I., Rakamaric, Z.: Evaluation of android malware detection based on system calls. In: Proceedings of the 2016 ACM on International Workshop on Security and Privacy Analytics, pp. 1–8 (2016)
3. Gupta, R., Sharma, K., Garg, R.K.: Covalent bond based android malware detection using permission and system call pairs. Comput. Mater. Continua **78**(3) (2024)
4. Handa, A., Sharma, A., Shukla, S.K.: Machine learning in cybersecurity: a review. Wiley Interdisc. Rev. Data Mining Knowl. Disc. **9**(4), e1306 (2019)
5. He, H., Yang, H., Mercaldo, F., Santone, A., Huang, P.: Isolation forest-voting fusion-multioutput: a stroke risk classification method based on the multidimensional output of abnormal sample detection. Comput. Methods Prog. Biomed., 108255 (2024)
6. Huang, P., et al.: Mamlformer: priori-experience guiding transformer network via manifold adversarial multi-modal learning for laryngeal histopathological grading. Inf. Fusion **108**, 102333 (2024)
7. Huang, P., et al.: La-vit: a network with transformers constrained by learned-parameter-free attention for interpretable grading in a new laryngeal histopathology image dataset. IEEE J. Biomed. Health Inf. (2024)
8. Laisin, M: Dynamic programming application of problem with optimal subsequence (2019)
9. Liu, K., Shengwei, X., Guoai, X., Zhang, M., Sun, D., Liu, H.: A review of android malware detection approaches based on machine learning. IEEE Access **8**, 124579–124607 (2020)
10. Mercaldo, F., Ciaramella, G., Santone, A., Martinelli, F.: Obfuscated mobile malware detection by means of dynamic analysis and explainable deep learning. In: Proceedings of the 18th International Conference on Availability, Reliability and Security, pp. 1–10 (2023)
11. Mercaldo, F., Zhou, X., Huang, P., Martinelli, F., Santone, A.: Machine learning for uterine cervix screening. In: 2022 IEEE 22nd International Conference on Bioinformatics and Bioengineering (BIBE), pp. 71–74. IEEE (2022)
12. Selvaraju, R.R., Cogswell, M., Das, A., Vedantam, R., Parikh, D., Batra, D.: Gradcam: visual explanations from deep networks via gradient-based localization. In: Proceedings of the IEEE International Conference on Computer Vision, pp. 618–626 (2017)
13. Surendran, R., Thomas, T., Emmanuel, S.: On existence of common malicious system call codes in android malware families. IEEE Trans. Reliab. **70**(1), 248–260 (2020)
14. Kumar, D., Radhamani, G., Vinod, P., Shojafar, M., Kumar, N., Conti, M.: Identification of Android malware using refined system calls. Concurr. Comput. Pract. Exp **31**, e5311 (2019)
15. Wang, H., et al.: Score-cam: score-weighted visual explanations for convolutional neural networks. In: Proceedings of the IEEE/CVF Conference on Computer Vision and Pattern Recognition Workshops, pp. 24–25 (2020)

16. Xiao, X., Xiao, X., Jiang, Y., Liu, X., Ye, R.: Identifying android malware with system call co-occurrence matrices. Trans. Emerg. Telecommun. Technol. **27**(5), 675–684 (2016)
17. Xiao, X., Zhang, S., Mercaldo, F., Hu, G., Sangaiah, A.K.: Android malware detection based on system call sequences and LSTM. Multimedia Tools Appl. **78**, 3979–3999 (2019)

A Portable Research Honeypot for Corporate Networks

Lorenzo Mirabella[1], Cinzia Bernardeschi[2], Giuseppe Lettieri[2]([⊠]),
Fabio Lucattini[1], and Salvatore Montanarella[1]

[1] Desys s.r.l, Centro Italmaco 1, Viareggio, Lucca, Italy
{l.mirabella,f.lucattini,s.montanarella}@desys.it
[2] Department of Information Engineering, University of Pisa, Pisa, Italy
{cinzia.bernardeschi,giuseppe.lettieri}@unipi.it

Abstract. The contemporary cybersecurity landscape faces an ongoing and dynamic threat environment, characterized by the persistent evolution of tactics employed by malicious actors. The detection and mitigation of these threats pose significant challenges, especially when dealing with individuals possessing intimate knowledge of an organization's security measures and vulnerabilities. Intrusion Detection Systems (IDS) play a crucial role in monitoring network traffic and systems for anomalies, providing alerts and defensive actions when suspicious activities are detected. While traditional IDS solutions exist, there is an increasing demand for adaptable and portable intrusion detection mechanisms. Honeypots, deceptive cybersecurity mechanisms designed to lure potential attackers, play a pivotal role in modern cyber-defense. By emulating vulnerable services, the honeypot captures data on the attacker's activities and diverts the attention away from the actual critical systems, enabling the enhancement of the overall network security. We describe the design, implementation and evaluation of a portable honeypot for intrusion detection in a corporate network, able to detect internal and external threats. Portability and platform-independency are ensured using Docker containers with a strong emphasis on security through the implementation of necessary measures to mitigate risk. The system adopts a microservices architecture and utilizes the Grafana stack for log collection, data visualization, and alert management. The study provides insights into security best practices and contributes to the ongoing efforts to strengthen cybersecurity defenses in an evolving threat landscape.

Keywords: honeypots · containers

1 Introduction

The contemporary cybersecurity landscape is characterized by a constant and evolving threat environment. Malicious actors employ a wide array of tactics, techniques, and procedures to breach network defenses and compromise data.

F. Martinelli and R. Rios (Eds.): STM 2024, LNCS 15235, pp. 128–139, 2025.
https://doi.org/10.1007/978-3-031-76371-7_9

Threats can be divided into two main classes: *External Threats* and *Internal Threats*. External Threats include attacks from external entities, such as cybercriminals who aim to infiltrate corporate networks, steal valuable data or install malware. Internal Threats, instead, are caused by individuals with authorized access to an organization's systems and data, but who misuse their privileges for malicious purposes. Internal threats can be particularly challenging to detect and mitigate, as they often involve individuals who possess intimate knowledge of an organization's security measures and vulnerabilities. This type of threat also includes attacks perpetrated by external entities but originating from an internal machine that has been taken over. Intrusion detection systems (IDS) play a crucial role in the cybersecurity context [1]. These systems serve as vigilant sentinels, continuously monitoring network traffic and systems. When anomalies or suspicious activities are detected, IDS can raise alerts and take defensive actions to mitigate potential threats. While traditional IDS solutions exist, there is a growing demand for portable and adaptable intrusion detection mechanisms [2–6]. Systems can be monitored by utilizing logs, statistics, and distributed traces, providing an opportunity to detect intrusion attempts, identify security flaws, and preemptively thwart attacks before they can cause significant damage. The concept of honeypot [7,8], a deceptive system designed to lure and engage potential attackers, takes many of these concerns into consideration and presents a unique approach.

The primary goal of this research is to design, implement, and evaluate an honeypot capable of effectively detecting a wide spectrum of threats. This includes mainly: (i) *Information Gathering*: The system must identify preliminary reconnaissance activities, such as port scanning, which are indicative of potential threats, whether originating from internal or external sources. (ii) *Exploitation of Vulnerabilities*: The system must be able to detect attempts to exploit vulnerabilities within applications or in the system. This includes scenarios where attackers may target web applications, potentially jeopardizing the organization's security. (iii) *Portability and Platform Independence*: Another key objective is to develop a portable solution without platform dependencies. This system should be deployable across different environments and architectures, ensuring flexibility and ease of installation. To provide insights into potential threats, the honeypot will log as much information as possible. The objectives for this task include: (i) *Data Collection*: Discerning valuable information regarding cyberattacks and their perpetrators. This data can be utilized to create methods for attributing attacks to particular sources, actors and to study their purposes. (ii) *Safe Data Handling*: Implementing security measures to protect the collected data, guaranteeing its integrity, confidentiality, and authenticity.

A honeypot can be deployed through various methods, but the proposed solution is purpose-built for installation within an organization's internal network. By following this approach, it can identify suspicious activities originating from both internal and external sources. In the event of the latter, this becomes particularly valuable as it signifies that potential threats have managed to circumvent the firewall or other security barriers. The system aims to become an integral part of the organization's security measures, ensuring early detection of threats.

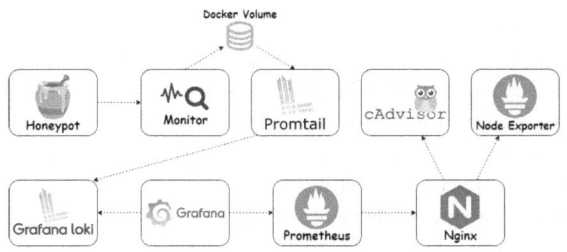

Fig. 1. Microservice architecture of the proposed system.

2 A Portable Honeypot

This section describes our prototype portable honeypot, designed for high interaction and suitable for corporate networks. The design leverages containers and virtual networks for portability, easy upgrades, and scalability. The architecture aims to enhance undetectability and reduce security risks despite high interaction and internal network deployment.

The system comprises nine containers, each with a specific role (Fig. 1). The core system includes two containers: the honeypot, hosting vulnerable services, and the monitor, which tracks activities. The honeypot contains a web application and an SSH service, while the monitor integrates monitoring tools, acting as a proxy for both services, and includes a Web Application Firewall (WAF) and an IDS.

Separating the honeypot and monitor into two containers minimizes honeypot fingerprinting, as the system interacting with the attacker shows no logging activity or monitoring tools, reducing suspicion.

Node Exporter [9] and cAdvisor [10] are used to monitor the host and Docker containers. Node Exporter exposes kernel and machine-level metrics, including CPU usage, memory, disk I/O, and network bandwidth. cAdvisor collects data on resource usage and performance of active containers, providing historical usage and network statistics.

Prometheus [11] gathers data from these exporters. Prometheus is an open-source monitoring system that uses labels for flexible analysis and supports Grafana [12] for visualization. It is known for high performance and easy integration into existing infrastructures.

The Grafana stack consists of Promtail [13], Loki [14], and Grafana [12]. Grafana is an open-source tool for querying, visualizing, and alerting on time-series data from various sources. Loki is a logging platform that stores and queries logs, using Promtail as a log collector. Promtail labels, transforms, and forwards logs to Loki, and queries can be visualized in Grafana.

The ninth container is an Nginx proxy managing requests to the exporters. The monitor writes logs to a Docker volume, which is also mounted on Promtail in read-only mode.

The Grafana stack was chosen for its simplicity, ease of installation, compatibility with various architectures, and strong community support, offering advantages over alternatives like ELK [15].

2.1 Monitoring and Alerting

The monitor container incorporates ModSecurity [16], Snort [17], and ssh-mitm [18] as monitoring and logging tools.

ModSecurity is a web application firewall (WAF) and HTTP intrusion detection tool aimed at enhancing web application security. It analyzes, records, and reacts to HTTP traffic in real-time without altering the application's code. Integrated with the reverse proxy Nginx via the ModSecurity-Nginx connector, it protects against attacks like SQL Injection, Cross-Site Scripting, and Local File Inclusion. In this setup, ModSecurity is in DetectionOnly mode, logging all requests and their details.

Snort operates as a network intrusion detection system (IDS), acting as a packet sniffer and logger with a set of updated detection rules. Using Snort within a container is challenging due to Docker's iptables rules, which prevent certain packets from reaching the bridge interface. This issue affects the detection of port scanning techniques. The port mirroring technique is used to forward packets from one interface to another, bypassing the host's iptables.

Ssh-mitm is a man-in-the-middle SSH server for security audits and malware analysis, supporting both password and public key authentication. It detects if a user can log in using public key authentication on the remote server and can redirect sessions to a honeypot if no agent is forwarded. It logs comprehensive SSH session information, including commands executed. Although the monitoring is transparent to attackers, the source IP shown during commands like netstat is the gateway's address, which could arouse suspicion but is not conclusive.

Grafana has been configured to provide a dashboard for each of the above monitoring tools, plus cAdvisor and Node Exporter. Grafana can send alerts triggered by configurable conditions.

2.2 Deployment

The honeypot container has a single network interface, e.g., eth0, with an IP like 198.168.0.2/24. The monitor container has two interfaces: eth1 with IP 192.168.0.3/24 on the same subnet as the honeypot, and eth0 on a different subnet, e.g., 192.168.1.3/24. This setup allows the monitor to intercept all traffic to the honeypot (Fig. 2). iptables routes traffic between eth0 and eth1.

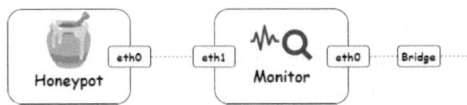

Fig. 2. The monitor container intercepts all traffic directed to the honeypot.

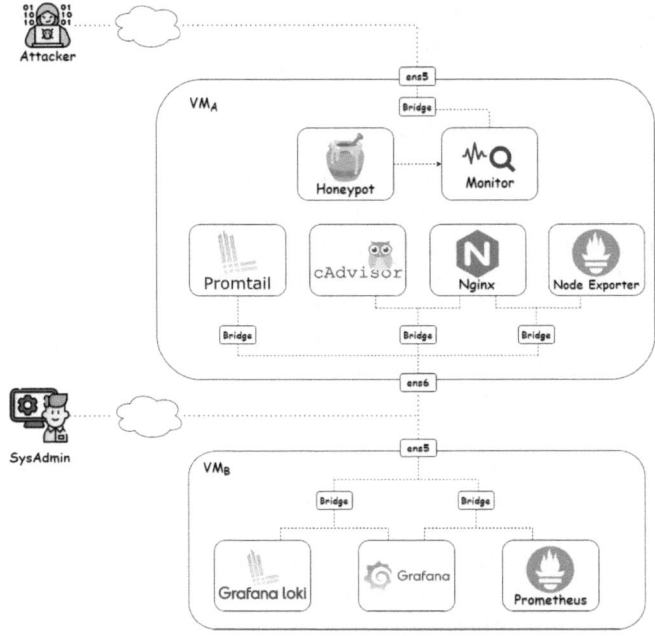

Fig. 3. Optimal deployment.

Docker Compose orchestrates the containers. Deployment can use one or more physical or virtual machines. The optimal setup uses two VMs: VMA (honeypot, monitor, cAdvisor, Node Exporter, Promtail) and VMB (Grafana Stack, Prometheus). This isolates attacker interactions to VMA, with administrators accessing VMB for dashboards.

In the ideal setup, VMA has two interfaces: ens5 (10.0.2.x/24) and ens6 (10.0.1.x/24). ens5, connected to the monitor's bridge, is reachable by attackers. ens6, on the same subnet as VMB's interface, is used for management, log streams to Loki, Prometheus requests, and SSH. This separation ensures complete isolation between malicious traffic and management traffic.

To enable external reachability to the machines, assigning a public address to each interface is necessary. However, this gives rise to a dilemma: it is not possible to reach both interfaces without specific routing rules. For instance, assume that the routing table is configured as in Table 1. This implies that if you ping the ens6 interface from the outside, the response traffic is directed to ens5 because it is the default gateway, resulting in unsuccessful pings. This issue does not arise when

Table 1. Default routing table for the deployment in Fig. 3.

Destination	Gateway	Genmask	Flags	Iface
0.0.0.0	10.0.2.1	0.0.0.0	UG	ens5
0.0.0.0	10.0.1.1	0.0.0.0	UG	ens6
10.0.1.0	0.0.0.0	255.255.255.0	U	ens6
10.0.2.0	0.0.0.0	255.255.255.0	U	ens5

pinging from VMB because there is a specific rule for 10.0.1.0/24. To address this, ens6 needs to be added as the default gateway for an IP address or subnet from which the administrator needs to connect to VMA via the management interface, using, for example, the following command:

```
route add -net 46.44.215.208/28 \
  gw 10.0.1.1 dev ens6
```

However, this introduces another limitation: that IP or subnet cannot contact the honeypot services for the same reason as before. Nevertheless, this is not a problem as those services are meant to be accessed by potential attackers, not administrators. Figure 3 shows the optimal configuration schema.

2.3   Security

Opting for the dual interface, as illustrated in Subsect. 2.2, offers several advantages. Firstly, it ensures that port 22 is not already occupied by the host machine itself. Additionally, conducting port scanning would reveal other services, although they can be filtered through a firewall. This guarantees with certainty that the attacker won't have visibility beyond the honeypot's content, providing assurance against interference. Furthermore, malicious traffic and management-related traffic would be completely segregated.

User Namespaces. Container isolation is achieved through control groups [19] and namespaces [20]. Namespaces isolate resources like pid, network, mount, and user namespaces. Docker uses all except user namespaces, which allow processes to have a customized view of user and group IDs. This means a root user in the container can be mapped to a non-root user on the host, removing the need for real root privileges in containers, thus reducing the risk of compromising container isolation. One cAdvisor container needs to open /dev/kmsg and does not use user namespaces.

SELinux. SELinux [21] (Security Enhanced Linux) is a Linux security architecture that gives administrators control over system access. It uses security policies to set rules for what can be accessed by applications, processes, and files. These rules ensure adherence to access criteria. If an attacker escapes the container and accesses the host, SELinux on the host minimizes their ability to cause damage.

Macvlan. Rather than linking container network interfaces to a Linux bridge on the Docker host, the macvlan [22] driver directly links the container interface to the Docker Host Ethernet interface. Each network connects to a distinct parent interface, creating a shared broadcast domain for containers. Communication with the default namespace IP address, like pinging the Docker host's eth0 from a container, is restricted. This kernel filtering enhances isolation and security.

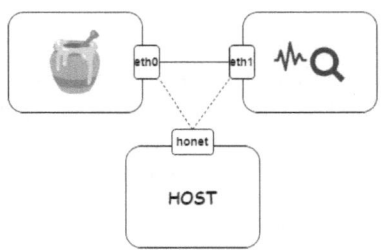

Fig. 4. The honet virtual network interface.

We created a dummy interface *honet* (see Fig. 4). A dummy interface, similar to the loopback interface, routes packets without transmitting them. In Docker Compose, the macvlan interface is attached to the dummy interface, functioning as a switch with two ports: one for the honeypot and one for the monitor. Frames cannot reach the host.

To validate, broadcast frames were sent using Python's Scapy from both the honeypot and the host. Monitoring the honet interface with tcpdump shows the frames, but they do not reach any other host interface, ensuring kernel-level isolation.

Resource Management. To mitigate the impact of DOS attacks, resource limits on CPUs and memory have been set on all containers. For example, if a fork bomb runs on the honeypot, control groups ensure the host remains unaffected. This is configured in the Docker Compose file.

To enhance reliability, especially during attacks, the Docker Compose file includes the option 'restart: always'. This ensures that containers automatically restart if they stop. If manually halted, they restart when the Docker daemon restarts or the container is manually started.

3   Tests

To assess the system's intrusion detection effectiveness, we conducted a series of tests culminating in a complex attack to determine if the collected data could discern the attacker's actions. The honeypot was also tested in a real-world scenario by deploying it in our University's data center.

3.1   Basic Tests

We simulated a brute-force SSH attack to guess the root password of the honeypot. Snort successfully detected the attack. Then, we attempted to connect to the honeypot and read /etc/passwd. The dashboard showed SSH connection attempts and session IDs. Selecting an ID allowed us to view the executed command and its result (Fig. 5).

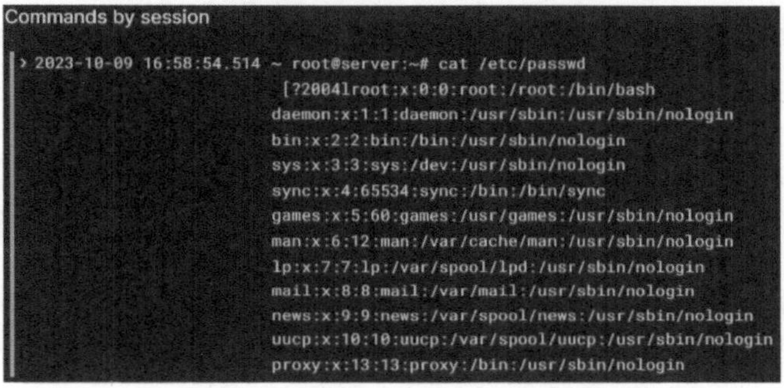

Fig. 5. Viewing attacker's actions through the ssh-proxy.

3.2   Simulation of a Complete Attack

A basic web app was deployed to simulate a complete attack involving SQL injection and XSS. The attack stages were: (i) Information gathering with nmap to find the web server. (ii) Blind SQL Injection on the login page to reveal username and password hashes. (iii) XSS to retrieve the admin's session cookie. (iv) Session Hijacking to gain admin access and execute commands.

The dashboard and data allowed us to reconstruct the attack. Snort detected port scanning, logging the source IP. ModSecurity displayed initial requests to the web server and subsequent requests made with Python's Requests library (Fig. 6). It also showed XSS, SQL Injection, and File Access Attempts (Fig. 7).

3.3   Real World Test

The honeypot was deployed on two virtual machines in the University of Pisa's data center, each with a public IP. Within 24 h, requests arrived globally, mainly from automated scanners.

The data was analyzed to assess monitoring and isolation effectiveness. During the week-long experiment, the host maintained stable parameters despite numerous attacks. SSH connections were few as port 22 was not used, and only two unsuccessful password attempts were made.

Top Attackers
ModSec Activity
Type of Attacks
1389
3
1385
1
Malicious Requests

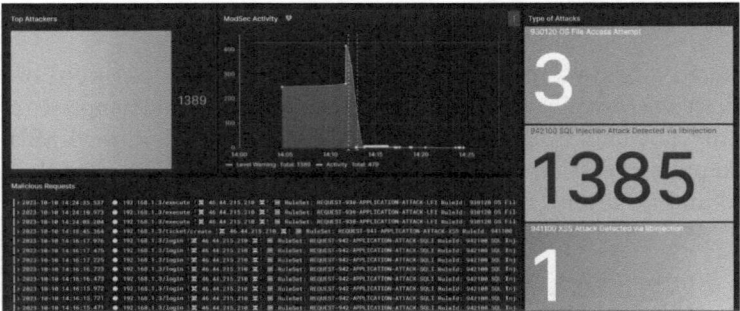

Fig. 6. Mod security dashboard during the web attack.

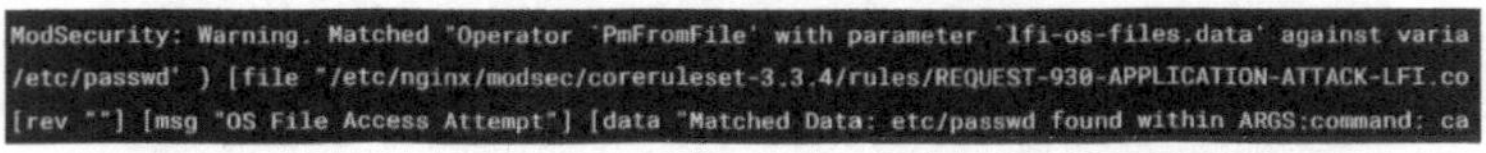

ModSecurity: Warning. Matched "Operator `PmFromFile' with parameter `lfi-os-files.data' against varia
/etc/passwd') [file "/etc/nginx/modsec/coreruleset-3.3.4/rules/REQUEST-930-APPLICATION-ATTACK-LFI.co
[rev ""] [msg "OS File Access Attempt"] [data "Matched Data: etc/passwd found within ARGS:command: ca

Fig. 7. Log of the commands in the final stage of the web attack.

For the web server, most requests were from automated scanners checking for vulnerabilities or exposed files. Snort detected several malicious packets and port scans. Both the host and containers showed no anomalies.

The 253 source IPs collected were from diverse locations worldwide, including India, Singapore, and the USA (Fig. 8). While no serious attack attempts were observed, valuable data was collected, and the solution was adequately tested.

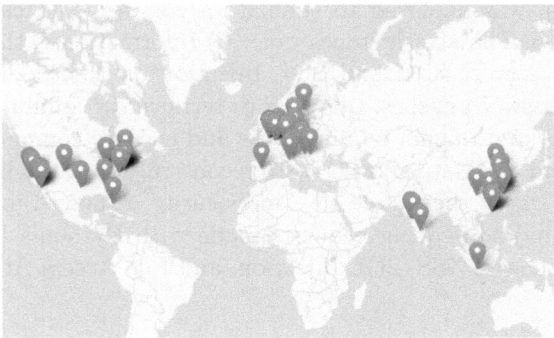

Fig. 8. Source IP map for the attacks during the real-world test.

4 Related Work

In recent years, several honeypot solutions have been introduced. Specifically, in 2023, two container-based alternatives closely resembling this work's proposal emerged [2,3]. Both solutions entail deploying open-source honeypots like

Cowrie [23], Dionaea [24], Glastopf [25], ADBHoney [26], and others. While currently offering a broader range of services compared to the presented solution, they rely on well-known projects that are familiar to attackers, significantly increasing the likelihood of honeypot detection. Furthermore, these solutions tend to overlook security concerns, lacking comprehensive security measures. An additional advantage of the proposed solution is the inclusion of host monitoring to detect potential anomalies. In essence, this approach represents an innovative strategy, offering insights to significantly enhance the security of these critical and vulnerable systems.

5 Conclusions and Future Work

The proposed solution addresses key drawbacks of honeypots by minimizing detectability and enhancing security.

Detectability is reduced by eliminating monitoring software, mitigating honeypot fingerprinting. Security measures, such as container isolation, virtual machines, user namespace, and SELinux, make it challenging for attackers to cause harm. The system, deployable via Docker containers, adopts a microservices architecture for easy maintenance, updates, scalability, and resilience. Complete undetectability, however, remains a problem. To address these limitation in future developments, a potential enhancement to the proposed system involves reimagining the monitor to achieve complete transparency. This revamped approach could intercept all traffic directed to the honeypot, silently decrypting encrypted HTTPS and SSH traffic using the known private keys, and subsequently subjecting it to analysis with the same tools. While implementing such a solution would be extremely challenging, its realization would effectively eliminate the current limitations, resulting in a theoretically more accurate and secure solution.

Acknowledgments. Work funded by the Italian Ministry of University and Research (MUR) in the framework of the FoReLab and CrossLab projects (Departments of Excellence).

References

1. Tidjon, L.N., Frappier, M., Mammar, A.: Intrusion detection systems: a cross-domain overview. IEEE Commun. Surv. Tutor. **21**(4), 3639–3681 (2019). https://doi.org/10.1109/COMST.2019.2922584
2. Priya, V.S.D., Chakkaravarthy, S.S.: Containerized cloud-based honeypot deception for tracking attackers. Sci. Rep. **13**(1), 1437 (2023). https://doi.org/10.1038/s41598-023-28613-0
3. Kyriakou, A., Sklavos, N.: Container-based honeypot deployment for the analysis of malicious activity. In: 2018 Global Information Infrastructure and Networking Symposium (GIIS) (2018)
4. Eftimie, S., Racuciu, C.: Honeypot system based on software containers. Sci. Bull. Mircea cel Batran Naval Acad. **19**(2), 582 (2016)

5. Gupta, C.: Honeykube: designing a honeypot using microservices-based architecture (2021)
6. Jonsson, M., Lysholm, A., Pham, S., Sannholm, B., Svanström, O., Valfridsson, J.: Developing a modular, high-interaction honeypot using container virtualization (2021)
7. Spitzner, L.: Honeypots: catching the insider threat. In: 19th Annual Computer Security Applications Conference, Proceedings, pp. 170–179 (2003). https://doi.org/10.1109/CSAC.2003.1254322
8. Mokube, I., Adams, M.: Honeypots: concepts, approaches, and challenges, In: Proceedings of the 45th Annual Southeast Regional Conference, ACM-SE 45, pp. 321–326. Association for Computing Machinery, New York (2007). https://doi.org/10.1145/1233341.1233399
9. Node Exporter. https://github.com/prometheus/node_exporter
10. cAdvisor. https://github.com/google/cadvisor/blob/master/docs/web.md
11. Pivotto, J., Brazil, B.: Prometheus: Up & Running. O'Reilly Media, Inc., Newton (2023)
12. Chakraborty, M., Kundan, A.P.: Grafana. In: Monitoring Cloud-Native Applications, pp. 187–240. Apress, Berkeley, CA (2021). https://doi.org/10.1007/978-1-4842-6888-9_6
13. G. Labs, Promtail documentation (2023). https://grafana.com/docs/loki/latest/send-data/promtail/
14. Bautista, E., Sukhija, N., Deng, S.: Shasta log aggregation, monitoring and alerting in hpc environments with grafana loki and servicenow. In: IEEE International Conference on Cluster Computing (CLUSTER) 2022, pp. 602–610 (2022). https://doi.org/10.1109/CLUSTER51413.2022.00079
15. Andreassen, O., Charrondière, C., De Dios Fuente, A.: Monitoring mixed-language applications with elastic search, logstash and kibana (ELK). WEPGF041 (2015). https://doi.org/10.18429/JACoW-ICALEPCS2015-WEPGF041. https://cds.cern.ch/record/2213498
16. Ristic, I.: Modsecurity handbook, Feisty Duck (2010)
17. Waleed, A., Jamali, A.F., Masood, A.: Which open-source ids? snort, suricata or zeek. Comput. Netw. **213**, 109116 (2022)
18. M. Kaiser, ssh-mitm offical documentation. https://docs.ssh-mitm.at/index.html
19. Cgroups - linux manual page. https://man7.org/linux/man-pages/man7/cgroups.7.html
20. Cheng, S.-T., Horng, G.-J., Hsu, C.-W., Su, Z.-Y.: Per-user network access control kernel module with secure multifactor authentication. J. Supercomput. **80**, 1–39 (2023)
21. What is selinux? https://www.redhat.com/it/topics/linux/what-is-selinux
22. Claassen, J., Koning, R., Grosso, P.: Linux containers networking: performance and scalability of kernel modules. In: NOMS 2016 - 2016 IEEE/IFIP Network Operations and Management Symposium, pp. 713–717 (2016). https://doi.org/10.1109/NOMS.2016.7502883
23. Cabral, W., Valli, C., Sikos, L., Wakeling, S.: Review and analysis of cowrie artefacts and their potential to be used deceptively. In: International Conference on Computational Science and Computational Intelligence (CSCI) 2019, pp. 166–171 (2019). https://doi.org/10.1109/CSCI49370.2019.00035
24. Ali, P.D., Kumar, T.G.: Malware capturing and detection in dionaea honeypot. In: 2017 Innovations in Power and Advanced Computing Technologies (i-PACT), pp. 1–5 (2017). https://doi.org/10.1109/IPACT.2017.8245158

25. Mphago, B., Mpoeleng, D., Masupe, S.: Deception in web application honeypots: case of glastopf. Int. J. Cyber-Secur. Digital Forensics **6**(4), 179–185 (2017)
26. Bythwood, W., Bentley, J., Vakilinia, I.: Analyses of automated malicious internet traffic using open-source honeypots. In: SoutheastCon 2023, pp. 68–75. IEEE (2023)

Unleashing AI in Ethical Hacking

Haitham S. Al-Sinani[1]($\boxtimes$)(iD), Chris J. Mitchell[2](iD), Nabil Sahli[3](iD),
and Mohamed Al-Siyabi[4]

[1] Department of Cybersecurity and Quality Assurance, Diwan of Royal Court,
Muscat, Oman
`hsssinani@diwan.gov.om`
[2] Department of Information Security, Royal Holloway, University of London,
Egham, Surrey TW20 0EX, UK
`C.Mitchell@rhul.ac.uk`
[3] Department of Computer Science, German University of Technology in Oman,
Muscat, Oman
`nabil.sahli@gutech.edu.om`
[4] Military Technological College, Muscat, Oman
`Mohamed.Al-Siyabi@mtc.edu.om`

Abstract. This paper explores the potential use of generative Artificial
Intelligence (GenAI) to enhance the effectiveness and efficiency of ethical hacking, and outlines a proof-of-concept implementation. It briefly
reviews the fundamentals of GenAI with a focus on ChatGPT, and then
summarises the concept and phases of ethical hacking. The paper also
critically assesses risks such as misuse of AI, data biases, and the danger
of over-dependence on technology, emphasising the importance of a collaborative human-AI partnership. The paper concludes with a discussion
of possible future directions, including use of AI in strengthening cyber
defences. This research contributes to the ongoing dialogue around the
ethical and innovative application of AI to bolster security.

Keywords: AI · Ethical Hacking · Generative AI · ChatGPT

1 Introduction

Ethical hacking, a key proactive security measure, traditionally requires a high
level of skill and constant updating of knowledge to work effectively. This paper
examines the use of ChatGPT, a state-of-the-art generative AI (GenAI) model,
to assist ethical hacking. The significance of this approach lies in the potential of GenAI to enhance the capabilities of ethical hackers, allowing for more
sophisticated and efficient security assessments and, ultimately, stronger defence
against cyber threats. In considering the adoption of GenAI in ethical hacking,
this paper not only contributes to current discourse but also aims to spark further research, fostering a more secure digital environment in an age where cyber
threats continue to evolve rapidly.

© The Author(s), under exclusive license to Springer Nature Switzerland AG 2025
F. Martinelli and R. Rios (Eds.): STM 2024, LNCS 15235, pp. 140–151, 2025.
https://doi.org/10.1007/978-3-031-76371-7_10

The remainder of the paper is organised as follows. Section 2 explores Chat-GPT and GenAI, and Sect. 3 highlights the ethical hacking landscape. Section 4 presents the GenAI-ethical hacking interoperation model, and Sect. 5 introduces a prototype implementation. Section 6 discusses the potential benefits and risks. Section 7 reviews related work, and, finally, Sect. 8 summarises the conclusions and outlines our plans for future work.

2   Generative AI and ChatGPT

The advent of GenAI, with models like ChatGPT[1] [4] being prominent, is a transformative shift in the AI landscape. These systems, moving beyond traditional AI's focus on pattern recognition and decision-making, excel in content creation, including text, images, and code. The ability to learn from extensive datasets and produce outputs that mimic human creativity is a major advance.

Central to this revolution is the GPT (Generative Pre-trained Transformer) architecture, the basis of models like ChatGPT. Developed by OpenAI, GPT models are built on deep learning techniques using *transformer* models, designed specifically for handling sequential data. These models undergo pre-training, where they learn from a wide array of Internet texts, followed by fine-tuning for specific tasks. This process enables models to grasp not just the structure of language but also its context, essential for generating human-like text.

Each iteration of ChatGPT has demonstrated enhanced sophistication, contextual understanding, and relevance of its outputs. Its primary function lies in interpreting user prompts and generating coherent, contextually appropriate responses. This versatility extends from conducting conversations to performing complex tasks, including coding, content creation, and, as we propose in this paper, ethical hacking. The GPT model family, including ChatGPT, owes much of its success to the transformer model, introduced in the 2017 paper: 'Attention is All You Need [11].' This architecture revolutionises sequence processing through attention mechanisms, enabling the model to focus on different parts of the input based on its relevance to the task.

As we explore the intersection of AI and cybersecurity, understanding Chat-GPT's foundational aspects becomes vital. Its generative nature, contextual sensitivity, and adaptive learning capacity can lead to innovative approaches in cybersecurity practices. Our focus will be on how these qualities of ChatGPT can be harnessed to revolutionise methodologies in ethical hacking, exploring the technical, ethical, and practical implications of this integration.

3   The Ethical Hacking Landscape

Ethical hacking, or penetration testing [10], involves applying hacking techniques to help organisations strengthen their security posture. It is a discipline that requires a blend of technical expertise, creativity, and a deep understanding

---

[1] https://openai.com/blog/chatgpt.

of potential threats. Unlike illegal hacking, ethical hacking is the authorised practice of bypassing system security to identify potential vulnerabilities and thereby help protect against real cyber attacks. Ethical hacking typically follows a structured approach involving five phases, as follows.

1. **Reconnaissance:** This involves gathering detailed information about the target system, including public data, network structures, IP address ranges, live hosts, and system configurations. The objective is to understand the target's environment and identify potential vulnerabilities.
2. **Scanning and Enumeration:** In this phase, ethical hackers use various tools to detect open ports, and exploitable system weaknesses, such as vulnerable services. The goal is to map out the target's network and identify entry points for potential attacks.
3. **Gaining Access:** During this stage, hackers leverage identified vulnerabilities to penetrate the target system. The aim is to gain an initial foothold, demonstrating how an attacker could potentially breach security measures and obtain unauthorised access.
4. **Maintaining and Elevating Access:** This stage involves researching and developing ways to gain elevated access and re-enter the system undetected. Techniques such as backdoors are implemented to enable long-term, potentially privileged access. Additionally, tactics like pivoting and lateral movement are employed to navigate through the network, accessing various systems to increase control and maintain persistence within the environment.
5. **Covering Tracks and Reporting:** This stage focuses on erasing any traces of the hacking activity to avoid detection and restoring the target system to its original state. Following this, ethical hackers prepare a comprehensive report detailing their findings and providing recommendations for improving the system's security posture.

While we have chosen to follow the *Penetration Testing Execution Standard* (PTES [2]) for its comprehensive and systematic approach, we acknowledge the existence of other reputable standards in the field. For instance, the *Open Source Security Testing Methodology Manual* (OSSTMM [8]) offers a robust framework for security testing, including physical and human security vectors. The *Information Systems Security Assessment Framework* (ISSAF) [1] provides a wide-ranging protocol for information security assessment, and NIST SP 800-115 [10] gives detailed guidelines for testing information systems, integrating well with risk management frameworks. Each of these standards has its strengths and can be suitable for different testing environments or organisational needs. Additionally, the OWASP (*Open Web Application Security Project*) Top 10 [9] serves as an essential reference for web application security.

4 Generative AI-Ethical Hacking Interoperation

We now propose a conceptual model for using the capabilities of GenAI to support ethical hackers across the five stages of ethical hacking. The model involves employing GenAI's advanced natural language processing and generation skills to enhance and streamline the tasks undertaken by pen-testers in each phase. We next summarise the proposed model's application in each stage of ethical hacking. We are currently undertaking a series of hands-on, research-driven experiments to empirically validate the approach.

4.1   Reconnaissance

The model is built on the assumption that GenAI can significantly enhance the reconnaissance phase of ethical hacking by automating collection and analysis of information. It can be programmed to gather data from multiple open-source intelligence (OSINT) resources, social media platforms, and public databases to create comprehensive profiles of a target system. By processing natural language, GenAI can sift through large volumes of text to identify useful information, such as potential entry points, system configurations, and network infrastructure details. GenAI could also be set to perform domain-specific reconnaissance, such as identifying tech stack details from developer forums or technical documentation, enabling a more targeted approach to vulnerability assessment. GenAI can create believable pretexts for social engineering, e.g. by generating messages that appear legitimate, to gather information without arousing suspicion.

4.2   Scanning and Enumeration

During this phase, GenAI interacts with network scanning tools and software to interpret outputs and suggest next steps. Via integration with existing network scanners and vulnerability assessment tools, it allows ethical hackers to ask questions in natural language about the target system's security posture. Additionally, GenAI's ability to learn from new data could be used to recognise novel patterns of vulnerabilities or calibrate scanning tools more effectively based on the system being assessed. GenAI interprets outputs from tools like Nmap and Wireshark, prioritises vulnerabilities, identifies anomalies, and cross-references findings with CVE databases to guide penetration testing.

4.3   Gaining Access

To help gain access, GenAI could aid ethical hackers by suggesting known exploitation techniques relevant to discovered vulnerabilities. It could simulate an attacker's reasoning by proposing attack vectors and generating proof-of-concept code snippets for exploiting vulnerabilities. GenAI could also provide insights into the latest exploit databases and frameworks, such as Metasploit modules, ensuring that ethical hackers are equipped with cutting-edge knowledge to identify and demonstrate risks effectively.

4.4   Maintaining and Elevating Access

While maintaining access is typically not a focus for ethical hackers beyond demonstration purposes, we argue that GenAI can still contribute by outlining how attackers could establish persistence. GenAI can offer guidance on elevating access as well as establishing persistent system access, all while remaining undetectable. It could generate hypothetical scenarios and suggest methods for creating backdoors or rootkits based on the vulnerabilities identified. This information can be valuable for understanding the risks associated with APTs (Advanced Persistent Threats) and formulating defences against them. In addition, GenAI can significantly assist ethical hackers by simulating the techniques of pivoting and lateral movement. This capability empowers the ethical hackers to comprehend and execute the essential strategies required to navigate through and expand their influence within a network.

4.5   Covering Tracks and Reporting

GenAI can assist in covering tracks and restoring the target system to its original state, crucial for avoiding detection during ethical hacking. It can identify traces left by attackers and propose effective methods for their removal, thus aiding in testing the robustness of a system's logging and monitoring capabilities. Furthermore, GenAI is likely to be adept at analysing system logs to detect forensic evidence of an attack and recommending strategies to conceal these traces, enhancing our knowledge of anti-forensic techniques. Additionally, we anticipate that GenAI will prove highly effective in documenting the ethical hacking process, summarising key findings, and generating comprehensive reports. These reports would typically include detailed descriptions of the techniques employed, vulnerabilities uncovered, and strategic recommendations for remediation.

5   A Proof-of-Concept

To validate the proposed approach, we conducted an experimental study in a controlled virtual environment. We assessed ChatGPT's effectiveness in aiding penetration testing, including testing Windows virtual machines with Kali Linux as the attack machine. A comprehensive account of the experiment is provided in a separate publicly available technical report [3]. However, we will include here the key details for the completeness of this paper.

5.1   Laboratory Setup

Physical Host. The experiment utilised a MacBook Pro with 16 GB RAM, a 2.8 GHz Quad-Core Intel Core i7 processor, and 1 TB of storage, providing sufficient computational capabilities for virtualisation.

Virtual Environment Configuration. The virtualisation of the network was achieved using VirtualBox 7, a reliable tool for creating and managing virtual machine environments. The setup included the following virtual machines (VMs).

1. **Kali Linux VM:** this machine functioned as the primary attack platform for conducting the penetration tests. It is equipped with the necessary tools and applications for ethical hacking.
2. **Windows VM:** this machine, running a 64-bit version of Windows Vista with a memory allocation of 512 MB, was the principal target for penetration testing within the experiment.
3. **Linux VM:** this machine, operating on a 64-bit Linux Debian system and allocated 512 MB of memory, was reserved for future testing and analysis.

The network configuration was established in a local NAT (Network Address Translation) network setup, allowing for seamless communication between the VMs and simulating a realistic network environment suitable to penetration testing and cybersecurity research.

5.2   Generative AI Tool

The experiment leveraged ChatGPT-4[2] for its advanced AI capabilities and efficient response time. The selection of ChatGPT-4 was primarily based on its prominent status as a leading GenAI tool, offering cutting-edge technology to enhance the ethical hacking process. It's crucial to note, however, that other GenAI tools are also available, e.g. Google's Bard[3] and GitHub's Co-Pilot[4], which could potentially be utilised in similar contexts. The methodologies and processes described are applicable to both the paid and free versions of Chat-GPT, with the paid version chosen for improved performance in this study.

5.3   Objective and Methodology

Our objective is to leverage ChatGPT's capabilities to assist in the ethical hacking process, aiming to gain unauthorised access to the target Windows VM. The experiment followed the five structured phases of ethical hacking (see Sect. 4), with ChatGPT's guidance integrated at each step, as follows.

Reconnaissance. In this paper, our emphasis is on active reconnaissance (recon), which necessitates engaging with the target to stimulate responses for observation. Therefore, during this phase, we have followed the steps listed below.

– As an integral part of the initial reconnaissance phase, our aim is to identify active machines within the target network in order to select our target. To

---

[2] https://openai.com/gpt-4.
[3] https://bard.google.com/.
[4] https://github.com/features/copilot/.

achieve this, we posed the following question to ChatGPT: "I'm currently in the initial stage of ethical hacking, known as Reconnaissance. Could you please provide a list of the top 4 commands I can use on my Kali machine to find out which devices are currently active on my local network?". As depicted in technical report [3], ChatGPT responded with a useful compilation of potential Kali terminal commands, including **nmap, netdiscover, and arp-scan,** along with examples on their utilisation.

- Next, we have transitioned to our Kali 'attack' machine and applied the recommendations provided by ChatGPT. As a result, we have successfully identified the active devices within the target network.
- To determine the IP address of our Kali 'attack' machine, we employed the 'hostname' command with the '-I' option.
- To make well-informed estimations regarding potential target machines, we can exclude both our Kali IP address and the standard default gateway. We can seek guidance from ChatGPT, wherein it analyses the 'arp-scan' command output, listing active network nodes, and the 'hostname' command output, specifying the Kali IP address. ChatGPT performs this analysis and offers educated insights via a question-answer chat communication.
- As a result of the analysis presented above, we can pinpoint the VMs with the IP addresses 192.168.1.6 and 192.168.1.7 as potential targets. This allows us to proceed to the second scanning stage.

Scanning and Enumeration. During this stage, ethical hackers use automated tools to scan the target system or network for vulnerabilities. This can include port scanning, vulnerability scanning, and more. In our specific scenario, the system that demands our scanning attention is the Windows machine identified by the IP address: '192.168.1.6'.

Analogous to preparing for a physical break-in, where determining the house address (reconnaissance) is the initial step, now that we have acquired the address (192.168.1.6), we can proceed to conduct a thorough inspection of the house to determine if any windows (ports) left open that may represent vulnerabilities.

We initiated this phase by consulting ChatGPT for a compilation of key commands suitable for gathering comprehensive information about a specific target (192.168.1.6) using our Kali machine. We explained to ChatGPT that our objective was to acquire extensive knowledge about this particular system to prepare for a forthcoming attack. ChatGPT thus provided us with a brief list of potential scanning commands, prominently featuring 'nmap -A -T4 192.168.1.6'. ChatGPT further clarified that the 'nmap -A -T4 192.168.1.6' command performs an aggressive scan, including OS detection, version detection, script scanning, and traceroute. The '-T4' option speeds up the scan with an aggressive timing template. Nmap, by default, scans the top 1,000 TCP ports; however, to scan all 65,535 ports on the target, we simply add the '-p-' option.

Subsequently, we employed the ChatGPT-recommended key command, 'nmap -A -T4 -p- 192.168.1.6' to perform an exhaustive scan of our target machine. The nmap scan results, unequivocally identifying our Windows target VM, are presented in our technical report [3]. We tasked ChatGPT with the analysis of these results and solicited suggestions for potential unauthorised access routes. ChatGPT highlighted that the nmap scan unveiled potential avenues for attack, most notably the open SMB (Server Message Block) ports 139 and 445, which may harbor vulnerabilities, including the infamous EternalBlue (MS17-010) exploit for remote code execution. We will then consult ChatGPT on exploiting the EternalBlue vulnerability, progressing to the next phase (see below).

Gaining Access. In this phase, we seek guidance from ChatGPT to gain access to the Windows VM with the IP address 192.168.1.6 using our Kali attack machine. To simplify the process, we have chosen to exploit the EternalBlue vulnerability via Metasploit. Our request to ChatGPT involves receiving instructions on utilising Metasploit on our Kali machine to execute the EternalBlue (MS17-010) attack after first confirming the system's vulnerability to this exploit. As depicted in our technical report [3], ChatGPT has provided a step-by-step guide. We begin by launching Metasploit with the 'msfconsole' command, and, then, proceed to search for the EternalBlue module using 'search eternalblue'. Next, we select the EternalBlue exploit module with 'use exploit/windows/smb/ms17_010_eternalblue', set the necessary options, including the target host IP address using 'set RHOSTS 192.168.1.6', and, optionally, configure the payload, which is set by default anyway. ChatGPT advises checking the target's vulnerability with the 'check' command, thereby confirming the system's susceptibility. Finally, to execute the exploit, we run the 'exploit' command, resulting in successful system ownership and system (root) access. For a visual representation of this step-by-step process, please refer to the publicly available technical report [3].

Maintaining and Elevating Access. In this stage of ethical hacking, our objective is to ensure we can re-enter the system in the future, ideally without being detected. Typically, achieving persistent access requires elevated privileges, often in the form of administrator or root access. As a result, we would typically turn to ChatGPT to assist us in elevating our access level. Fortunately, in the previous stage, we successfully exploited the 'EternalBlue' vulnerability, granting us system access (the highest level of privileged access possible). With this in mind, we consulted ChatGPT for guidance on maintaining persistent access. As shown in our technical report [3], ChatGPT provided a list of recommendations for establishing persistent access. These include creating backdoors, utilising scripts for persistence, manipulating services or scheduled tasks, DLL hijacking, and modifying registry keys. For simplicity, we requested a step-by-step guide from ChatGPT on creating a basic backdoor by adding a new user account with administrative privileges. ChatGPT did indeed offer a detailed guide, which involves creating a new user account using the command: 'execute -f cmd.exe -c -H -i -a "/c net user newusername password /add"', adding this user to the administrators group using the command: 'execute -f cmd.exe -c -H -i -a "/c net localgroup administrators newusername /add"', and verifying the new user's addition, such as through the command: 'execute -f cmd.exe -c -H -i -a "/c net localgroup administrators"'. Following ChatGPT's instructions meticulously, we confirmed the successful addition of the new user to the admin group. Subsequently, we also tested this by restarting the Windows target machine and successfully confirmed our ability to log in using the newly created user through the standard Windows login procedure.

Covering Tracks and Documentation. This phase comprises two parts:

– **covering our tracks,** which involves erasing or minimising evidence of our activities within the target system, crucial to avoid detection and maintain the system as close to its original state as possible; and
– **documentation,** which involves creating the pen-test report, a topic discussed later.

In the first part, aiming to remain undetected, we asked ChatGPT for a detailed guide on effectively covering our tracks. As shown in our technical report [3], ChatGPT provided a list of actions to achieve this, including:

– the removal of the newly added user account (Haitham) using the command: 'meterpreter execute -f cmd.exe -c -H -i -a "/c net user Haitham /delete"';
– clearing system logs with the command: 'meterpreter > clearev';
– deleting any files created or downloaded onto the target system;
– uninstalling any software or tools;
– resetting system settings;
– removing scheduled tasks for persistence;
– flushing DNS and ARP cache with 'ipconfig /flushdns' and 'arp -d *' to eliminate network activity traces; and, finally,
– gracefully closing the 'Meterpreter' session using the 'exit' command.

While we have implemented some of these recommendations, such as clearing system logs, it's worth noting ChatGPT's caution that clearing logs can raise suspicion in real-world scenarios and might not always be advisable.

As for the second part, the documentation part, it is crucial for ethical hackers to produce a comprehensive and thorough report for each penetration testing assignment. Therefore, we asked ChatGPT to assist us in composing a detailed report for our penetration test (simulation) assignment using all the information that ChatGPT already knows about from our chat-based communication. As shown in our technical report [3], ChatGPT first responded with a template that we can use to structure our report, along with guidance on what to include in each section. Since providing the template was not satisfactory, we asked ChatGPT again to write a comprehensive and detailed report for this penetration testing assignment using the recommended template and the information we have discussed in our chat, incorporating as much detail as feasible and providing supporting evidence where relevant. This time around, ChatGPT responded with a well-written and accurate penetration test report, including writing the 'Executive Summary, 'Introduction', 'Methodology', 'Findings and Results', 'Attack Narrative', 'Conclusions and Recommendations', as well as suggestions for 'Appendices' . In subsequent questions to ChatGPT, we further tweaked and improved the ChatGPT-produced report, including specifying the target organisation, time period, and the date.

6   Discussion: Benefits and Potential Risks

We have introduced a methodical approach to using the capabilities of GenAI to augment each stage of ethical hacking, from initial data gathering to final reporting. The model aims to support provision of a thorough and efficient evaluation of security vulnerabilities. It assists in the detailed and rapid identification of potential threats, contributes to the development of social engineering scenarios, and aids in formulating defensive measures through simulated attacks. Implementing this model could lead to more streamlined security assessments and potentially improve the training of cybersecurity professionals, thereby equipping them with enhanced skills to identify and address evolving cyber threats.

However, adoption of this model is not without challenges, and requires careful consideration of the accompanying risks and ethical implications. The potential for misuse by adversaries, the risk of over-dependence on automated systems at the expense of human expertise, and the AI's current limitations in processing ambiguous or deceptive information present substantial concerns. Additionally, biases inherent in AI algorithms and the lag in regulatory frameworks to address such rapidly advancing technologies pose further complications. These factors highlight the need for a balanced integration of AI systems, such as ChatGPT, into ethical hacking practices, ensuring that it complements human judgment rather than replacing it, and is governed by a robust ethical framework to maximise its potential while safeguarding against misuse.

7   Related Work

The intersection of AI and cybersecurity is a vibrant area of research, with studies ranging from AI's role in detecting intrusions to its use in aiding offensive security including ethical hacking. Foundational work by Handa et al. [6] has showcased the value of machine learning in network intrusion detection. The rise of sophisticated language models like GPT-3, introduced by Brown et al. [4], has heralded new research possibilities, such as exploring AI's use in crafting realistic phishing attacks. Contemporary studies, for example Gupta et al. [5], look at the dual role of AI, showing how it could both be employed for cyberattacks and harnessed for cyber defence and ethical guidance. Furthermore, a recent practical study by Harrison et al. [7] shows how advances in AI's deep learning algorithms enhance acoustic side-channel attacks against keyboards, achieving groundbreaking keystroke classification accuracies with common devices and apps like smartphones and Zoom.

This paper seeks to expand on these discussions, offering an exploration of GenAI's role across all stages of ethical hacking—a topic that remains underexplored in the existing literature, necessitating a deeper investigation.

8   Conclusions and Future Directions

This paper has discussed the potential role of GenAI as a tool for ethical hacking, and introduced a framework for its application. We have highlighted the collaborative capabilities of human expertise paired with AI's computational power for cybersecurity. The paper also outlined a proof-of-concept implementation; the initial results of which clearly show that ChatGPT, a state of the art GenAI model, is an effective and impactful tool in the field of ethical hacking. The paper sets the agenda for future empirical research to further validate the assertions underlying our GenAI-ethical hacking interoperation model.

Ongoing research includes a series of hands-on, research-driven experiments aimed at both substantiating the proposed model and refining it to encompass a wider array of hacking domains. This entails increasing the model's scope across other security disciplines including wireless security, privilege escalation, protection against the OWASP Top 10 (web[5] and mobile[6]) vulnerabilities, and mobile app security. By conducting these experiments, we aim to continuously evolve the model to address the ever-changing landscape of cyber threats, ensuring its effectiveness against increasingly sophisticated future attack vectors.

References

1. Open Information Systems Security Group (OISSG): Information Systems Security Assessment Framework (ISSAF) (2006). https://www.untrustednetwork.net/files/issaf0.2.1.pdf
2. Penetration Testing Execution Standard (PTES). PTES Technical Guidelines (2014). https://www.pentest-standard.org/index.php/Main_Page
3. Al-Sinani, H., Mitchell, C.: Unleashing AI in ethical hacking: a preliminary experimental study. Technical report, Royal Holloway, University of London (2024). https://pure.royalholloway.ac.uk/files/58692091/TechReport_UnleashingAIinEthicalHacking.pdf
4. Brown, T.B., et al.: Language models are few-shot learners. Adv. Neural Inf. Process. Syst. **33**, 1877–1901 (2020). https://arxiv.org/abs/2005.14165
5. Gupta, M., Akiri, C., Aryal, K., Parker, E., Praharaj, L.: From chatgpt to threatgpt: impact of generative AI in cybersecurity and privacy. IEEE Access, **11**, 80218–80245 (2023). https://doi.org/10.1109/ACCESS.2023.3300381
6. Handa, A., Sharma, A., Shukla, S.K.: Machine learning in cybersecurity: a review. WIREs Data Min. Knowl. Disc. **9**(4), e1306 (2019). https://doi.org/10.1002/WIDM.1306

---

[5] https://owasp.org/www-project-top-ten/.

[6] https://owasp.org/www-project-mobile-top-10/

7. Harrison, J., Toreini, E., Mehrnezhad, M.: A practical deep learning-based acoustic side channel attack on keyboards. In: IEEE European Symposium on Security and Privacy, EuroS&P 2023 - Workshops, Delft, Netherlands, 3–7 July 2023, pp. 270–280. IEEE (2023). https://doi.org/10.1109/EUROSPW59978.2023.00034
8. Institute for Security and Open Methodologies (ISECOM): Open Source Security Testing Methodology Manual (OSSTMM) (2020). https://www.isecom.org/OSSTMM.3.pdf
9. OWASP (Open Worldwide Application Security Project). OWASP top ten (2021). https://owasp.org/www-project-top-ten
10. Swanson, M., Bartol, N., Sabato, J., Hash, J., Graffo, L.: Technical guide to information security testing and assessment (NIST SP 800-115). Special Publication 800-115, National Institute of Standards and Technology (2008). https://csrc.nist.gov/publications/detail/sp/800-115/final
11. Vaswani, A., et al.: Attention is all you need. In: Guyon, I.,et al. (eds.) Advances in Neural Information Processing Systems 30: Annual Conference on Neural Information Processing Systems 2017, December 4-9, pp. 5998–6008. Long Beach, CA, USA. (2017). https://proceedings.neurips.cc/paper/2017/hash/3f5ee243547dee91fbd053c1c4a845aa-Abstract.html

Author Index

© The Editor(s) (if applicable) and The Author(s), under exclusive license
to Springer Nature Switzerland AG 2025
F. Martinelli and R. Rios (Eds.): STM 2024, LNCS 15235, p. 153, 2025.
https://doi.org/10.1007/978-3-031-76371-7